Intelligent knowledge based systems in electrical power engineering

Intelligent knowledge based systems in electrical power engineering

J.R. McDonald

Centre for Electrical Power Engineering
University of Strathclyde
Scotland, UK

G.M. Burt

Centre for Electrical Power Engineering
University of Strathclyde
Scotland, UK

J.S. Zielinski

Department of Informatics
University of Lodz
Poland

and

S.D.J. McArthur

Centre for Electrical Power Engineering
University of Strathclyde
Scotland, UK

SPRINGER-SCIENCE+BUSINESS MEDIA, B.V

First edition 1997

Originally published by Chapman & Hall in 1997
Softcover reprint of the hardcover 1st edition 1997
Typeset in 10/12 Palatino by Florencetype Ltd, Stoodleigh, Devon

ISBN 978-1-4613-7934-8 ISBN 978-1-4615-6387-7 (eBook)
DOI 10.1007/978-1-4615-6387-7

A catalogue record for this book is available from the British Library

∞ Printed on permanent acid-free text paper, manufactured in accordance with ANSI/NISO Z39.48–1992 and ANSI/NISO Z39.48–1984 (Permanence of Paper).

Contents

List of contributors xi

Introduction 1

Part One
A General Overview of Intelligent Knowledge Based systems

1 An introduction to intelligent knowledge based systems 5
Jerzy S. Zielinski and Stephen D.J. McArthur
1.1 Computers and artificial intelligence 5
1.1.1 A short history 5
1.2 Current trends in applied AI 7
1.3 Definition and terminology of IKBSs 8
1.3.1 Definitions 8
1.3.2 Terminology 8
1.4 The architecture Of IKBSs 9
1.5 The implementation cycle of IKBSs 11
References 11

2 Intelligent knowledge based system development 13
Jerzy S. Zielinski and Stephen D.J. McArthur
2.1 Historical review 13
2.1.1 General information 13
2.1.2 IKBS tools 13
2.1.3 Applications of IKBSs 14
2.1.4 IKBS generations 15
2.2 IKBSs in technology 15
2.3 Opportunities and bottlenecks in IKBS development 16
2.4 Discussion of costs and benefits 17
References 17

Further reading 19

Part Two
The Application of Intelligent Knowledge Based Systems in Electrical Power Engineering

3 Comments on contemporary power systems and other engineering systems 27
Jerzy S. Zielinski
3.1 Main characteristics of electrical power systems 27
3.2 Problems in electrical power system operation 28
3.3 Computer applications in power systems 29
3.3.1 A short history 29
3.3.2 Some remarks on hardware and software 29
3.3.3 IKBSs 30
3.4 Some remarks on the relationships between phenomena in power systems and other fields of engineering 31
3.5 Conclusions 33
References 33

4 Forecasting and planning in power systems 35
U. George Knight
4.1 Introduction 35
4.2 Planning tasks in the extension of power systems 35
4.3 Forecasting tasks in the extension of power systems 36
4.4 The evolution of techniques for system planning and forecasting 37
4.5 The characteristics of system planning and forecasting tasks in relation to IKBSs 38
4.5.1 Algorithmic, human and expert system capabilities 39
4.5.2 The characteristics of planning and forecasting tasks in relation to IKBSs 40
4.5.3 Frequency of study 42
4.5.4 Retention of expertise 42
4.6 Areas of research and development 44
4.7 Descriptions of some proposed and actual applications of IKBSs in generation expansion planning 45
4.7.1 A decision tree and IKBS model 46
4.7.2 A combined dynamic programming, IKBSs and fuzzy sets approach 49
4.8 Descriptions of some proposed and actual applications of IKBSs in transmission network planning 55

4.8.1 An IKBS for the preliminary design of power transmission networks 56
4.8.2 Development of an IKBS for long-term planning of power transmission networks 58
4.9 Descriptions of some proposed and actual applications of IKBSs in distribution network planning 60
4.9.1 EHV/HV substation location and HV network design 61
4.9.2 Connection of new customers 62
4.10 Descriptions of some proposed and actual IKBSs in power system analysis 64
4.10.1 Interactive load flow for power system planning 65
4.10.2 Support to load flow calculations 65
4.10.3 Support for transient stability studies 66
4.11 Demand forecasting 70
4.12 Trends and conclusions 71
References 72

5 Design of electrical plant and systems **75**
Stefano Massucco
5.1 Introduction 75
5.1.1 General overview 75
5.1.2 Reasons for the application of IKBSs to design and planning 76
5.2 The state of the art in IKBS applications for the design of electrical plant and systems 78
5.2.1 Introduction 78
5.2.2 Applications to large systems 79
5.2.3 Applications to systems of limited dimensions 80
5.2.4 Applications to the design of specific subsystems or equipment 82
5.3 An IKBS application: electrical auxiliary system design in a power station 83
5.3.1 Problem formulation 83
5.3.2 The IKBS approach: building the knowledge base and design procedure 87
5.4 Conclusions 95
Acknowledgements 96
References 97

6 Intelligent knowledge based systems in condition monitoring **99**
Barclay Weir
6.1 Introduction 99

6.2 Types of electrical machines 100
6.3 Condition monitoring 102
6.4 IKBSs in condition monitoring 105
6.5 Applications 106
6.6 Some Successful applications 107
6.6.1 TURBOMAC 108
6.6.2 GenAID 108
6.6.3 GEMS 110
6.6.4 TOGA 111
6.6.5 MOTORMON 114
6.7 Summary 116
References 117

7 Alarm processing **119**
David J. Young and James R. McDonald
7.1 Introduction 119
7.2 The history of alarm processing 120
7.2.1 Message routing 120
7.2.2 Message prioritization 121
7.2.3 Advanced alarm-processing techniques 121
7.2.4 Review of previous work 123
7.3 The practical problems of alarm processing 127
7.3.1 Speed of operation 127
7.3.2 Limited scope of telemetry 128
7.3.3 Missing alarm messages 128
7.3.4 Unreliable time ordering of alarms 129
7.3.5 Arrival rate of alarms is not constant 129
7.3.6 Multiple simultaneous events 130
7.4 Development of an alarm-processing expert system 130
7.4.1 Setting initial targets 132
7.4.2 Identification of knowledge and data requirements for the IKBS 133
7.4.3 Identification of existing data and knowledge 135
7.4.4 Knowledge acquisition 135
7.4.5 Identification of approach and IKBS techniques 137
7.4.6 Identification of language or tools for IKBS construction 138
7.4.7 Knowledge base design 139
7.4.8 User interface design 140
7.4.9 IKBS construction – knowledge base and user interface 140
7.4.10 Prototyping 141
7.4.11 Testing 141
7.4.12 Installation and integration 142

7.5 A case study: APEX 143
7.5.1 Details of APEX construction 144
7.5.2 Example of APEX operation 146
7.5.3 Performance 151
7.6 Conclusion 152
References 152

8 Event and fault diagnosis 155
Graeme M. Burt and James R. McDonald
8.1 Introduction to the problem of fault diagnosis 155
8.1.1 Events 155
8.1.2 Reasons for fault diagnosis 156
8.1.3 Basis for diagnosis 156
8.1.4 Reasons for a fault diagnosis system 157
8.1.5 IKBS-based fault diagnosis 158
8.2 Review of IKBSs for network fault diagnosis 158
8.2.1 The 'automatic system trouble analysis' system 158
8.2.2 IKBSs based on ASTA ideas 159
8.2.3 A distributed approach 160
8.2.4 Fault diagnosis using object-oriented features 161
8.2.5 Localized fault diagnosis 162
8.2.6 Applications where SCADA data is lacking 163
8.2.7 A system suited to parallel implementation 164
8.2.8 A system for distribution feeders 164
8.2.9 CRAFT – customer restoration and fault testing 165
8.2.10 SEPT 165
8.2.11 An IKBS developed by Amantegui *et al.* 166
8.2.12 Further comments 166
8.3 The development of an IKBS for fault diagnosis: RESPONDD 167
8.3.1 Introduction 167
8.3.2 Structure of the diagnosis: its results 168
8.3.3 Structure of the diagnosis: its mechanics 170
8.3.4 An initial prototype: wait-and-see 171
8.3.5 Hypothesizer strategy 173
8.3.6 Hypothesizer gut reaction 174
8.3.7 Evolution of hypothesizing strategies for central diagnosis 174
8.3.8 Dealing with uncertainty 177
8.3.9 Network data requirements 179
8.3.10 Qualitative simulation 181
8.3.11 Modelling knowledge 184
8.4 Demonstration of the developed system 187
8.4.1 Case study 187

8.4.2 Hypothesizer diagnosis 189
8.5 Final remarks 190
References 192

9 Other applications in power system operation and control 195
Jerzy S. Zielinski
9.1 Voltage/reactive power control 195
9.2 Emergency control and restoration 196
9.3 Real-time processing 198
9.4 The testing of equipment 199
9.4.1 Transformers 199
9.4.2 Protection relays 200
9.4.3 Substations 200
9.5 Management systems 200
9.6 Electrical and other industries 201
References 201

10 Future trends in intelligent systems in power systems 206
Alan Moyes
10.1 Introduction 206
10.2 The importance of knowledge 206
10.2.1 Knowledge acquisition 207
10.2.2 Model-based reasoning 209
10.2.3 Case-based reasoning 210
10.2.4 Data mining 210
10.3 Integrated solutions for complex problems 211
10.3.1 Control and monitoring architecture 211
10.3.2 Integrated IKBSs 214
10.4 Conclusions 215
References 216

Index 219

Contributors

Graeme M. Burt
Centre for Electrical Power Engineering,
Department of Electrical and Electronic Engineering,
University of Strathclyde,
Royal College Building,
204 George Street,
Glasgow G1 1XW,
UK

U. George Knight
Department of Electrical and Electronic Engineering,
Imperial College of Science, Technology and Medicine,
Exhibition Road,
London SW7 2BT,
UK

Stefano Massucco
EPSL,
Electrical Engineering Department,
University of Genoa,
Via al' Opera Pia, 11a,
I–16145 Genoa,
Italy

Stephen D.J. McArthur
Centre for Electrical Power Engineering,
Department of Electrical and Electronic Engineering,
University of Strathclyde,
Royal College Building,
204 George Street,
Glasgow G1 1XW,
UK

James R. McDonald
Centre for Electrical Power Engineering,

Department of Electrical and Electronic Engineering,
University of Strathclyde,
Royal College Building,
204 George Street,
Glasgow G1 1XW,
UK

Alan Moyes
Centre for Electrical Power Engineering,
Department of Electrical and Electronic Engineering,
University of Strathclyde,
Royal College Building,
204 George Street,
Glasgow G1 1XW,
UK

Barclay Weir
Technology Division,
ScottishPower plc,
47 Hawbank Road,
East Kilbride G74 5EG,
UK

David J. Young
Scottish Nuclear,
Electrical and Control Design Department,
3 Redwood Crescent,
Peel Park,
East Kilbride G74 5PR,
UK

Jerzy S. Zielinski
Department of Informatics,
University of Lodz,
39 Rewolucji 1905r Str.,
90–214 Lodz,
Poland

Introduction

Technological advances in the field of computer science resulted in the development of new branches of intellectual activity, and one very important discipline resulting from this technology was that of artificial intelligence (AI). Nearly 20 years of research in this area (prior to 1965) was without success. However the development of the first expert system (ES), called MYCIN, broke this impasse. This was the first application to effectively use intelligent knowledge based system (IKBS)[1] techniques.

Computers have been used to solve a great number of practical problems in complex systems, and electrical power engineering is an area where computers have been extensively used. Both hardware and software achievements have been applied to power system problems. The first IKBS applied to power engineering was implemented in a nuclear power plant at the end of the 1970s, while the beginning of the 1980s saw the first applications of IKBSs to power systems. Following this, the number of papers devoted to IKBSs for power system applications grew rapidly. It is impossible to prepare a list of all these publications, and for that reason only some of these will be presented in this book with references to other relevant work.

This book consists of two main parts: Part One, 'A General Overview of Intelligent knowledge based Systems' (by J.S. Zielinski and S.D.J. McArthur) and Part Two, 'The Application of Intelligent knowledge based Systems in Electrical Power Engineering' (by a number of authors).

The first part consists of two chapters and a further reading list. The first chapter, 'An Introduction to Intelligent knowledge based Systems', presents a short history of AI development, definitions of IKBSs and IKBS terminology, and the architecture and building process for IKBSs.

[1] This is the preferred terminology within this book, and will be used instead of 'expert system', unless it is inappropriate to do so.

The second chapter, 'Intelligent knowledge based System Development', includes a historical review, some remarks on IKBS technology, the opportunities and bottlenecks in IKBS development, and a discussion of the costs and benefits of IKBS applications.

The second part of the book is composed of eight chapters written by a group of authors. Chapter 3 (J.S. Zielinski), 'Comments on Contemporary Power Systems and Other Engineering Systems', characterizes electrical power systems, presenting operational problems and computer applications in power systems. At the end of this chapter the author discusses the relationships between power system phenomena and other fields of engineering, as well as a basis for transferring IKBS achievements from the power system domain to other industries.

Chapter 4 (U.G. Knight), 'Forecasting and Planning in Power Systems', discusses the application of IKBSs to the domain of forecasting and planning in electrical power systems.

Chapter 5 (S. Massucco), 'Design of Electrical Plant and Systems', discusses the problems and solutions to the design problem for electrical plant and systems.

Chapter 6 (B. Weir), 'Intelligent Knowledge Based Systems in Condition Monitoring', covers the area of condition monitoring by introductory discussion and by examples of actual working systems.

Chapter 7 (D.J. Young and J.R. McDonald), 'Alarm Processing', gives a description of the problem of alarm processing for electrical utilities, and gives a detailed description of one particular example of an IKBS to resolve these problems.

Chapter 8 (G.M. Burt and J.R. McDonald), 'Event and Fault Diagnosis', gives an overall evaluation of the many attempts at producing a working IKBS-based solution to the problem of event and fault diagnosis in the electrical power systems domain. A description of a working system is also given in detail.

Chapter 9 (J.S. Zielinski), 'Other Applications in Power System Operation and Control', gives a valuable supplement to the more detailed specific chapters by introducing some additional application areas for IKBS solutions.

Finally, the book is concluded by an assessment of future directions in this domain through Chapter 10 (A. Moyes), 'Future Trends in Intelligent Systems in Power Systems'.

Part One

A General Overview of Intelligent knowledge based Systems

1

An introduction to intelligent knowledge based systems

Jerzy S. Zielinski and Stephen D.J. McArthur

Developments in computer science raise a number of philosophical as well as practical issues. One of these issues concerns the relationship between contemporary results and the original theoretical objectives, while another relates to the use of computer tools in research and practical applications. The aim of this book is to address these issues taking into account the techniques of artificial intelligence (AI) and its 'offspring' – intelligent knowledge based systems (IKBSs).

AI has fascinated people for several centuries [1]. However, modern research into AI began about 40 years ago as a result of the introduction of the first generation of digital computers. Results from AI research can be traced alongside the achievements of computers, indicating that AI and computer development are in parallel and are strongly linked. These links are evident in the definition of AI, which, according to Winston [2], '. . . is the study of ideas that enable computers to be intelligent'. Moreover he states that AI has two central goals '. . . to make computers more useful' and ' . . . to understand the principles that make intelligence possible' [2].

1.1 COMPUTERS AND ARTIFICIAL INTELLIGENCE

1.1.1 A short history

The introduction of the first electronic digital computer, ENIAC, caused not only enthusiasm but also unrealistic expectations in the intellectual community. The following quotation [3] is indicative of these attitudes: 'An electrical brain can impair the most solid base of our hitherto existing social valuation logic-intellect'. This opinion explains why the advent of computers influenced the development of AI in the early 1950s.

One can observe the following two main directions in AI research [4]:

1. the building of a physical device or a device simulation on a digital computer;
2. symbolic representations for building formal structures (models) capable of being solved by computers.

W.S. McCulloch and W.H. Pitt's work, presented in their paper published in 1943 [5], can be regarded as the first significant work in AI. Their neural-net model, called an M-P unit, was synchronous (using simple logical switches) quite unlike real neurons. Ten years later, F. Rosenblatt modified this net and called it a 'perceptron'. The perceptron could be trained to classify certain types of patterns as either similar or distinct.

In 1956, Newell and Simon solved simple puzzles and proved theorems in prepositional calculus by the second approach involving a symbolic representation and computer programming.

Both approaches were able to solve some easy problems. However it was impossible to extrapolate these results to the real world because increasingly complex problems needed exponentially growing computing time.

By 1970, brain simulation by neural networks was losing favour, while the second approach had full control of AI research funds. In seeking to decrease the neural-network role, Minsky and Papert analysed the limitations of the one-layer perceptron. (Their results, however, could not be applied to other nets, e.g. multilayer systems.)

The lack of results from practical applications prompted E. Feigenbaum and J. Lederberg's group to direct their AI research into the exploration of a real-world problem [1]: how do scientists confronted with a problem solve it? The result was the development of DENDRAL in the mid 1960s, the first IKBS to analyse the composition of organic molecules.

The history of AI development is very interesting considering the intellectual activity of scientists [1,4–16] concerned with related problems. For example, the 'common-sense knowledge problem' using the symbolic representation approach dominated AI development from 1975. The assumption that the common-sense knowledge problem must be solvable, since humans can solve it, blocked all progress in theoretical AI for the decade. The reason for this impasse was that human 'know-how' does not necessarily imply the existence of procedural rules, but rather knowledge of what to do in special cases [4].

This obstacle to the symbolic representation concept diverted AI research towards connectionism (i.e. neural-net research). Consequently the Rumelhart and McClelland book, *Parallel Distributed Processing* (1986), sold rapidly.

1.2 CURRENT TRENDS IN APPLIED AI

A number of techniques are currently popular within the area of applied AI. Each technique employs a different emulation of 'intelligent' behaviour and is applied to problems which complement its strengths. The key techniques presently in use are: neural networks; genetic algorithms; fuzzy systems; intelligent knowledge based systems.

The history of neural-network research was covered in section 1.1.1. In terms of practical applications, due to the learning capabilities of neural networks they are often employed where a problem requires estimation, prediction or classification. Neural networks do not store knowledge or expertise explicitly but implicitly characterize behaviour through a learning process [17].

A common class of problem, within AI applications, is where the solution is found through some type of search strategy. Genetic algorithms are favoured for search problems which require the identification of a globally optimal solution. They are based on the natural evolution process and operate by combining the best solutions at each stage into the next generation of solutions. This process is iterative until the criteria for an optimal solution are met. The popularity of the genetic algorithm approach can be attributed to the fact that optimal solutions can be found without the search becoming trapped in local minima or maxima. Additionally, genetic algorithms can be applied to problems with a large number of variables to be considered [18].

It is recognized that the analysis of data can be subjective. To allow for this situation fuzzy systems, using fuzzy mathematics, have been employed. Such systems allow an element of subjectiveness, or vagueness, to be associated with data. As a simple example, a temperature measurement may be manipulated by a fuzzy system through converting the actual temperature into one of the following qualitative values: very cold; cold; medium; hot; very hot. A useful application of such fuzzy systems is decision making under uncertain or hypothetical situations [19].

An IKBS differs from the other AI systems discussed in that it is an attempt to explicitly embody expertise and knowledge within software. The remainder of this book concentrates on the structure and design of IKBS plus their application within the realm of electrical power engineering.

1.3 DEFINITION AND TERMINOLOGY OF IKBSs

1.3.1 Definitions

In the definitions of IKBSs one can observe the following two approaches:

1. definitions involving the characteristics of an IKBS, mostly following from its architecture [20,21];
2. definitions involving an IKBS's function [22,23].

The first approach will not be considered now since the IKBS architecture will be developed in section 1.4.

A number of IKBS definitions can be found in the Further Reading section which follow the second approach, two examples of which are:

> An artificial intelligence system created to solve problems in a particular domain is called an 'expert system' [22].
> An ES is a computer application that solves complicated problems that would otherwise require extensive expertise. To do so, it simulates the human reasoning process by applying specific knowledge and inferences [23].

Comparing the above definitions one can see that the first includes the second, nevertheless the second gives more detail concerning the IKBS architecture.

1.3.2 Terminology

IKBS terminology results from two main sources: from computer science and from AI. The following are some of the most important terms in this book.

Domain knowledge – expert knowledge in the application area of the IKBS.
IKBS building tool – programming platform and language used for building an IKBS.
Explanation facility – means of explaining the reasoning that led to a conclusion.
General purpose knowledge engineering language – programming language used in IKBS building, containing procedures valid for different domains and systems.
Inference engine – the part of the IKBS that applies a reasoning strategy and generates solutions based on the rules in the knowledge base.
Interpreter – a part of the inference engine which resolves how to use the domain knowledge.
Knowledge base – the part of the IKBS containing the human expertise expressed through some appropriate mechanism (e.g. rules and facts).

Knowledge engineer – a person who interacts with the expert to elicit the domain knowledge.

Knowledge engineering – the process of acquiring specific domain knowledge and building it into the knowledge base [23,24].

Knowledge representation – the processes of building a knowledge structure and the structure's description [24].

Metaknowledge – knowledge an IKBS has about its internal knowledge.

User interface – the part of the IKBS which facilitates the communication process between the user and the rest of the IKBS.

1.4 THE ARCHITECTURE OF IKBSs

The majority view is that the IKBS architecture consists of the following parts: a knowledge base, an inference engine, an explanation facility and a user interface. Electrical power engineering applications need a much more developed IKBS structure which contains all the above parts with additional modules, as presented in Figure 1.1. A more detailed consideration of the functions of each part of an IKBS follows.

The **knowledge base** (KB) and the inference engine are essential to the IKBS. The KB stores the basic facts, procedural rules and heuristics of the system, and determines the expert capability of the system. The knowledge representation schemes in particular IKBSs vary, and these influence the design of the inference engine, knowledge updating and explanation processes, as well as the efficiency of the IKBS.

The **inference engine** enables the IKBS to respond to varying situations and to infer new knowledge from existing knowledge. If we were to try to store all possible combinations of knowledge in a particular domain we would reach a very large number of elements (a combinatorial explosion [23]). For this reason, compiled knowledge (high-level knowledge) is used in most IKBSs. The inference engine infers new knowledge from that in the knowledge base using a reasoning method (or paradigm). Most inference engine paradigms are search strategies and are based on one of two fundamental concepts: backward or forward chaining. The backward-chaining approach starts from the desired goals and goes backward to identify the requisite conditions. The forward-chaining approach starts with known conditions and goes towards the desired goal. According to Rolston [23], 'the selection of the inference paradigm, with consideration for combinatorial explosion, strongly influences the overall performance of the ES'.

The **knowledge update** facility allows the stored knowledge to be modified in accordance with changes to and expansion of the domain knowledge. The updating process can take one of the following three basic forms:

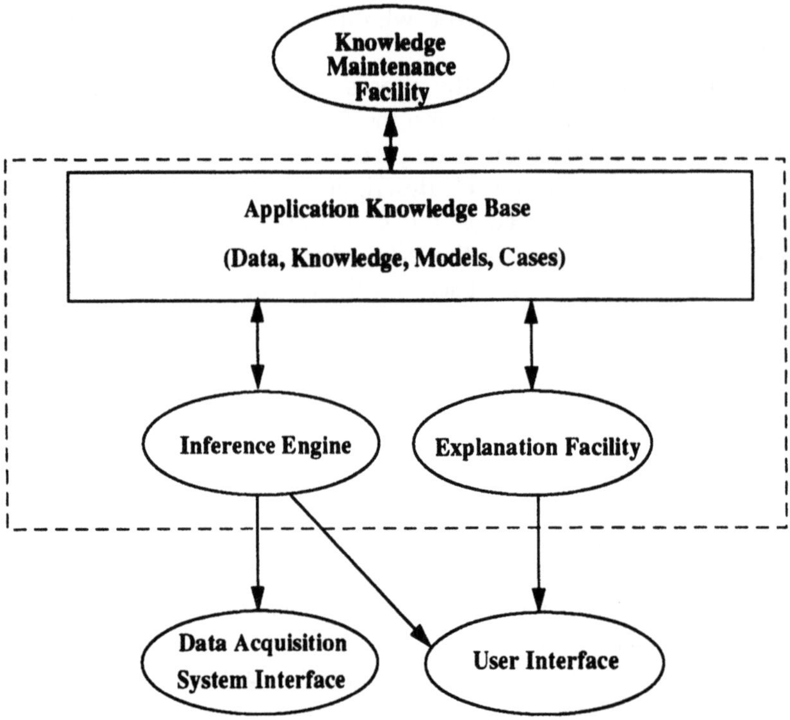

Figure 1.1 Structure of an expert system.

1. manual knowledge updated by a knowledge engineer interpreting the domain expert information;
2. direct updating of the knowledge without the knowledge engineer;
3. machine learning (self-updating).

An **explanation system** allows the user to view the reasoning process of the IKBS – a facility which is very important for establishing the credibility of the IKBS. The explanation system is also a useful training tool for a new user.

The **knowledge engineer** plays a very important role in acquiring knowledge from the domain expert and other sources. The knowledge engineer must transform the acquired knowledge into the convention of the knowledge representation system. The knowledge acquisition process may be the most difficult process in building an IKBS.

The remaining elements in Figure 1.1 are necessary for linking the IKBS to its environment, and belong to the relevant data acquisition sources. For example, the supervisory control and data acquisition (SCADA) systems or energy management systems (EMSs) of modern power systems [25,26].

1.5 THE IMPLEMENTATION CYCLE OF IKBSs

The design and implementation cycle of IKBSs differs somewhat from that of other software systems. In essence this can be attributed to there being two distinct parts of an IKBS: the knowledge base; the software which manipulates the knowledge and performs the reasoning. Both parts must be developed and validated in a structured fashion. As a result, the key stages of the IKBS implementation cycle are:

1. specification of the problem
2. knowledge analysis
3. design
4. implementation
5. operational evaluation and refinement.

Specification of the problem will result in the requirements of the IKBS being identified. This covers its purpose and functionality as well as consideration of performance issues. Through problem specification the requisite knowledge (and knowledge sources) plus data requirements of the IKBS will be identified.

The **knowledge analysis** stage is where the embedded expertise, to be realized within the IKBS's knowledge base, is elicited. It may come from diverse sources such as human experts, technical manuals and textbooks. The knowledge must be formalized into an appropriate representation and then validated (for example, by further analysis from human experts). This becomes an iterative process where each knowledge validation step further refines the knowledge base.

The **design** stage is the planning and design of the software which performs the reasoning process and manipulates the knowledge base. This entails selecting the IKBS tools and software languages to be used. Furthermore, input and output data requirements need to be defined along with the appropriate user interface functions.

Following the design process, **implementation** of the IKBS takes place. This is the realization of the data and knowledge bases, the inference engine, user interface etc. This stage includes the coding of the data acquisition functions which interface with the relevant data sources of the IKBS.

Operational evaluation and refinement is a validation process applied to the complete IKBS. It covers its on-line functionality and encompasses the knowledge and the software manipulating it. As with the knowledge validation stage, this process is iterative and refines both knowledge and software in step with on-line experience of the IKBS.

REFERENCES

1. McCorduck, P. (1988) Artificial intelligence: an apercu. *Daedalus, Journal of the American Academy of Arts and Sciences*, **117**(1), 65–83.

2. Winston, P.H. (1984) *Artificial Intelligence*. Addison-Wesley.
3. Zielinski, J.S. (1991) Some remarks on artificial intelligence and computers, in *Proceedings of Seminar on 'Expert Systems and Integrated Case-Tools'*, Lodz, September, pp. 30–38.
4. Dreyfus, H.L. and Dreyfus, S.E. (1988) Making a mind versus modelling the brain: artificial intelligence back at a branch point. *Daedalus, Journal of the American Academy of Arts and Sciences*, **117**(1), 15–43.
5. Cowan, J.D. and Sharp, D.J. (1988) Neural nets and artificial intelligence. *Daedalus, Journal of the American Academy of Arts and Sciences*, **117**(1), 85–121.
6. Dennett, D.C. (1988) When philosophers encounter artificial intelligence. *Daedalus, Journal of the American Academy of Arts and Sciences*, **117**(1), 283–95.
7. Hillis, W.D. (1988) Intelligence as an emergent behaviour; or the songs of Eden. *Daedalus, Journal of the American Academy of Arts and Sciences*, **117**(1), 175–89.
8. Hurlbert, A. and Poggio, T. (1988) Making machines (and artificial intelligence) see. *Daedalus, Journal of the American Academy of Arts and Sciences*, **117**(1), 213–39.
9. McCarthy, J. (1988) Mathematical logic in artificial intelligence. *Daedalus, Journal of the American Academy of Arts and Sciences*, **117**(1), 297–311.
10. Papert, S. (1988) One AI or many? *Daedalus, Journal of the American Academy of Arts and Sciences*, **117**(1), 1–14.
11. Putnam, H. (1988) Much ado about not very much. *Daedalus, Journal of the American Academy of Arts and Sciences*, **117**(1), 269–81.
12. Reeke, G.N. Jr and Edelman, G.M. (1988) Real brains and artificial intelligence. *Daedalus, Journal of the American Academy of Arts and Sciences*, **117**(1), 143–73.
13. David, J.-M., Krivine, J-P. and Simmons, R. (eds) (1993) *Second Generation Expert Systems*, Springer-Verlag.
14. Hamscher, W., Console, L. and de Kleer, J. (eds) (1992) *Readings in Model Based Diagnosis*, Morgan Kaufman.
15. Kelly, J. (1993) *Artificial Intelligence – a Modern Myth*, Ellis Horwood.
16. Kirschen, D., Bann, J. *et al.* (1995) Expert systems integration: issues and solutions. *IEEE PICA*.
17. Taylor, J.G. (1996) *Neural Networks and Their Applications*, Wiley.
18. Mitchell, M. (1996) *An Introduction to Genetic Algorithms*, MIT.
19. Terano, T., Asai, K. and Sugeno M. (1994) *Applied Fuzzy Systems*, Academic Press.
20. Bubnicki, Z. (1990) *Introduction to Expert Systems*, (in Polish). PWN, Warsaw.
21. Waterman, D.A. (1986) *A Guide to Expert Systems*, Addison-Wesley.
22. Levine, R.I., Drang, D.E. and Edelson, B. (1988) *A Comprehensive Guide to AI and Expert Systems Using Turbo Pascal*, McGraw-Hill Book Company.
23. Rolston, D.W. (1988) *Principles of Artificial Intelligence and Expert Systems Development*, McGraw-Hill Book Company.
24. Wielinga, B.J., Schreiber, A.Th. and Breuker, J.A. (1992) KADS: a modelling approach to knowledge engineering. *Knowledge Acquisition*, **4**(1), 5–53.
25. McArthur, S.D.J., McDonald, J.R., Bell, S.C. *et al.* (1995) The extension of corporate communications networks to realise intelligent data analysis: a case study providing protection system performance analysis, in *CIGRE Symposium on Integrated Control and Communication Systems*, Helsinki, August, paper 600–05.
26. Bann, J., Irisarri, G., Kirschen, D. *et al.* (1996) Integration of artificial intelligence applications in the EMS – issues and solutions. *IEEE Transactions on Power Systems*, **11**(1), 475–82.

2

Intelligent knowledge based system development

Jerzy S. Zielinski and Stephen D.J. McArthur

2.1 HISTORICAL REVIEW

2.1.1 General information

The first IKBS to be developed was DENDRAL in 1965, which analysed the composition of organic molecules. In the same year, two other IKBSs were also developed: MACSYMA and HEARSAY, which performed mathematical analysis and natural language interpretation respectively. The development in 1972 of MYCIN, an expert system for the diagnosis of blood diseases, was an important point in IKBS history. With the removal of MYCIN's knowledge from the knowledge base, a tool called EMYCIN (empty MYCIN) was developed. In the 1970s a number of other IKBSs appeared, for example: PROSPECTOR for mineral exploration; OPS5, a tool for IKBS building; RI for the configuration of DEC computer equipment.

In the early stages of IKBS history, IKBSs were developed on mainframe computers which imposed limitations on the widespread use of IKBSs. The age of the personal computer facilitated the rapid growth in the number of IKBSs. Now IKBSs are deliverable on all types of hardware platforms.

As a result of the interest in IKBSs, applications range from engineering through science and technology to medicine, law and finance [1].

2.1.2 IKBS tools

The IKBS development process is not easy and the use of appropriate tools is very important. One of the groups of tools are programming languages, and LISP is one of the most frequently used for IKBS development. The other main symbolic processing language is Prolog. Nonsymbolic languages such as C (and more recently, C++) are also used in IKBS development. The advantages of using C include: faster program execution than the LISP equivalent on general-purpose hardware; easy inclusion of numerical computations, which are important in a number of power-system-based IKBS applications.

It is also important to mention that, in order to improve the efficiency of IKBS development and operation, specialized LISP machines (AI workstations) have been developed for the execution of LISP.

Another method of developing intelligent knowledge based systems is to use a **shell**. An IKBS shell is essentially an IKBS with the knowledge stripped out of it, such as EMYCIN which was produced from MYCIN [2]. Shells provide facilities to implement a knowledge base and hence tailor the IKBS for a specific application. Such IKBS shells come in a variety of formats, offering variable amounts of design flexibility, and are suitable for a wide variety of hardware platforms [2–5]. IKBS shells aim to facilitate IKBS development without the complete control and inferencing methods needing to be designed.

2.1.3 Applications of IKBSs

Intelligent knowledge based systems are being applied in a wide variety of fields, as was stated in section 2.1.1. A number of publications have attempted to assess the number of applications and the preferred techniques within this technology [6,7]. However, the field of IKBSs is evolving at a rate which quickly renders such statistics out of date.

Moving from the application domain to the specific tasks of IKBSs shows an equally wide spectrum of activity. Nevertheless, some of the most frequently observed tasks include [7,8]:

1. design
2. diagnosis
3. decision-making support
4. control
5. planning.

These are generalized categories and for each industry sector (or application domain) more specific examples can be found. When considering electrical power engineering, IKBSs are applied to a number of problems and tasks [9]. These include: alarm processing; fault diagnosis; system restoration; security assessment; reactive power and voltage control; switching operation; load flow planning; transient stability problems; unit commitment; operator training; maintenance scheduling; substation automation; plant monitoring.

2.1.4 IKBS generations

In the beginning most IKBSs were rule-based in nature. These are referred to as shallow reasoning systems [10], and are called **conventional** or **first generation** systems. Shallow reasoning systems tend to suffer from the following limitations [10]:

1. a lack of robustness (they cannot solve unexpected cases);
2. representation of the available knowledge is difficult due to the limitation of rules;
3. they are not easily maintained in the case of large knowledge bases.

Second-generation IKBS, using model-based reasoning or reasoning from first principles, attempt to alleviate these limitations. According to Console *et al.* [11] second-generation IKBSs 'tend to have automated knowledge acquisition and learning capabilities, multiple levels of abstraction, multiple knowledge representation and generic task architectures'. This allows for a number of possible architectures for these IKBSs [11,12].

2.2 IKBSs IN TECHNOLOGY

When studying IKBS publications (e.g. Motoda [7]), it is easy to see the following application areas of IKBSs: interpretation, prediction, diagnosis, design, planning, monitoring, debugging, repair, instruction and control. These areas are also the most important fields of interest in the technology world, which indicates the very widespread application IKBSs have in industry. Waterman's book [13] presents 98 IKBSs developed for industrial application before 1986, but now the number of applications is considerable and difficult to determine.

Ishii and Hayami [14] present a list of 121 IKBSs applied in Japanese companies to aid engineers in different technical processes. Some of the specific technologies in which IKBSs have been applied are presented, for example, in references 15–18.

Industrial IKBSs can be either off-line or on-line systems. According to Johanssen and Alty [19], the off-line systems will support the design and maintenance personnel, while the on-line systems will support plant operators. These different classes of people not only need different types of knowledge but also use different knowledge sources. The off-line IKBSs, without special timing requirements, may use technical documents, literature, handbooks and knowledge about operations personnel (taking into account ergonomics), as their knowledge sources. The online IKBSs are time critical because, as support systems, they are required to operate in parallel with the dynamic industrial process. The speed of the process will influence the complexity of the IKBS development.

IKBSs applied in industry also have other specific properties. Johanssen and Alty [19] consider the following.

1. IKBSs need to be integrated with supervisory and control systems, and should support symbolic knowledge as well as numeric computations.
2. IKBSs may not be of use in every industrial activity since, for example, existing supervisory and control systems are based upon engineering methodologies and IKBSs can only support them and not replace them.

2.3 OPPORTUNITIES AND BOTTLENECKS IN IKBS DEVELOPMENT

As has been shown (section 2.2), contemporary IKBSs are applied to a wide range of functions. This wide variety of applications results from the following characteristics of IKBSs [2]:

1. extensive specific knowledge from the application domains of interest;
2. effective application of search techniques;
3. support for knowledge based analysis;
4. capacity to infer new knowledge from existing knowledge;
5. the power of symbolic processing;
6. an ability to explain its own reasoning.

In addition to the above, the following characteristics should be included [20].

1. The knowledge base is separate from the inference mechanism.
2. The development process of a system may be continuous.
3. The output of an IKBS is advisory.

When developing an IKBS it is necessary to remember that the quality of the IKBS is dependent on the accuracy of the acquired knowledge. The knowledge acquisition process is one of the key sources of problems in IKBS development and, according to [7], the following bottlenecks in IKBS implementation must be borne in mind:

1. incomplete hardware and software;
2. lack of knowledge engineers;
3. an absence of knowledge on how to develop and apply IKBS technology;
4. difficulty in receiving support from specialists.

2.4 DISCUSSION OF COSTS AND BENEFITS

It is not easy to consider in detail the costs and benefits of any computer information system, and this applies to IKBSs also. Nevertheless it is possible to collect some remarks devoted to these problems.

In general, the building and implementation of an IKBS will cost more than another computer information system dedicated to the same tasks (assuming that the computer information system can realize these tasks). This is due to the complexity of the knowledge acquisition process, as well as the loops in the IKBS life cycle (section 1.4).

In order to avoid the big losses associated with breakdown in an IKBS's implementation, a six-month rule is proposed in [17]. According to this, any company building an IKBS should assume the possible loss of six worker–months of expenses.

In spite of the costs of building an IKBS, their operation costs are lower (there being no need to ask the expert). This perhaps accounts for the rapid growth of IKBS applications in human activities, which is the best proof of their user acceptance (e.g. references 1,7,8,21–24). In general, the benefits following a successful IKBS implementation are a result of its capability to solve complex problems (within appropriate timescales) that other automated systems could not deal with.

In addition, it is very interesting to note the following information (presented in van de Riet [21]).

1. The growth of the IKBS market in the USA and Canada was $300 million in 1986 and $1 billion in 1990.
2. PROSPECTOR discovered a molybdenum deposit with the ultimate value probably exceeding $100 million.
3. Other IKBSs such as R1, DENDRAL, CADUCEUS etc., are now solving problems, some of which are insoluble with 'classical' approaches, in a number of companies all over the world.

REFERENCES

1. Milne, R. and Montgomery A. (eds) (1994) *Applications and Innovations in Expert Systems II (Proceedings of Expert Systems 94)*, SGES Publications.
2. Mayers, W. (1986) Introduction to expert systems. *IEEE Expert*, Spring, 100–108.
3. Harmon, P. and Sawyer B. (1990) *Creating Expert Systems*, John Wiley.
4. Citrenbaum, R., Geissman, J.R. and Schultz, R. (1987) Selecting a shell. *AI Expert*, September, 30–39.
5. Gevarter, W.B. (1987) The nature and evaluation of commercial expert system building tool. *Computer*, May, 24–41.
6. JIPDEC (1987) Results of survey on trends in expert systems in Japan. *Future Generation Computer Systems*, **3** (1), 17–36.
7. Motoda, H. (1990) The current status of expert system development and related technologies in Japan. *IEEE Expert*, August, 3–11.

8. Rada, R. (1990) Expert systems in UK. *IEEE Expert*, August, 12–17.
9. Liu, C.-C., Ma, T.-K., Liou K.-L. and Tsai, M.-S. (1993) Practical use of expert systems in power systems, in *Fourth International Symposium on Expert System Applications to Power Systems*, Melbourne, Australia, pp. 1–12.
10. Wang, X. and Dillon, T. (1991) A second generation expert system for fault diagnosis, in *Third Symposium on Expert Systems Applications to Power Systems*, Tokyo, April, pp. 751–56.
11. Console, L., Portinale, L. and Dupre, D.T. (1991) Focussing abductive diagnosis. *AI Communications*, **4**(2/3), 88–97.
12. Dague, P. (1995) Qualitative Reasoning: a survey of techniques and applications. *AI Communications*, **8**(3/4), 119–92.
13. Waterman, D.A. (1986) *A Guide to Expert Systems*. Addison-Wesley.
14. Ishii, K. and Hayami, S. (1988) Expert systems in Japan. *IEEE Expert*, Summer, 69–74.
15. Chang, K.-H. and Wee, W.G. (1988) A knowledge based planning system for mechanical assembly using robots, *IEEE Expert*, Spring, 18–30.
16. Davis, K. (1988) Diagnostic expert system for the B1B. *IEEE AES Magazine*, April, 20–25.
17. Ho, T.-L., Bayles, R.A. and Sieger, E.R. (1988) Aircraft VSCF generator expert system. *IEEE AES Magazine*, April, 6–13.
18. Scoping (1988) Estimation manpower requirements for expert systems project. *EXP*, Summer.
19. Johanssen, G. and Alty, J.L. (1991) Knowledge engineering for industrial expert systems. *Automatica*, **27**(1), 97–114.
20. Eliot, L.B. (1990) AI Conferences: the image is real. *IEEE Expert*, February, 78–79.
21. van de Riet, R.P. (1987) Problems with expert systems? *Future Generation Computer Systems*, **3**(1), 11–16.
22. Fauquembergue, P., Kezunovic, M. *et al.* (1995) Intelligent system applications to protections, control and monitoring within substations. *CIGRE Symposium on Integrated, Control and Communication Systems, Helsinki*, August, paper 600–06.
23. *Proceedings of Expert Systems 96 (ES96)*. Cambridge, UK, 16–18 December 1996.
24. Mohammed, O.A. and Tomsovic, K. (eds) (1996) *Proceedings of the International Conference on Intelligent System Applications to Power Systems (ISAP'96)*, IEEE.

Further reading

Aarts, E. (1989) *Simulated Annealing and Boltzmann Machines*, John Wiley.

Abdelguerfi, M. and Lavington, S. (1995) *Emerging Trends in Database and Knowledge Base Machines: The Application of Parallel Architectures to Smart Systems*, IEEE.

Adelman, L. (1992) *Evaluating Decision Support and Expert System Technology*, John Wiley.

Adey, R.A. (1991) *Artificial Intelligence in Engineering Design*, vol. 12, Computational Mechanics Publications, Southampton.

Adey, R.A. and Sriram, D. (eds) (1987) *Applications of Artificial Intelligence to Engineering Problems*, Computational Mechanics Publications, Southampton.

Adey, R.A., Rzevski, G. and Sunol, A.K. (eds) (1996) *Applications of Artificial Intelligence in Engineering XI*, Computational Mechanics Publications, Southampton.

Allwood, R.J. (1989) *Techniques and Applications of Expert Systems in the Construction Industry*, John Wiley.

Alty, J.L. and Johannsen, G. (1987) knowledge based dialogue for dynamic systems, in *10th World Congress on Automatic Control*, IFAC, Munich, July, vol. 7, pp. 358–67.

Anderson, D. (1989) *Artificial Intelligence and Intelligent Systems*, John Wiley.

Arbib, M.A. (1989) *The Metaphorical Brain 2*, John Wiley.

Aslett, M.J. (ed.) (1991) *A knowledge based Approach to Software Development*, Elsevier Science Publishers.

Ayel, M. and Laurent, J.P. (1991) *Validation, Verification and Tests of Knowledge Based Systems*, John Wiley.

Badard, R. (1987) Integrated process knowledge systems, in *10th World Congress on Automatic Control*, IFAC, Munich, July, vol. 6, pp. 314–318.

Banerji, R.B. (ed.) (1990) *Formal Techniques in Artificial Intelligence*, Elsevier Science Publishers.

Barbel, R. (1991) *Bones*, Simon & Schuster International Group.

Beardon, C. (ed.) (1989) *Artificial Intelligence Terminology*, John Wiley.

Beerel, A. (1991) *Expert Systems*, Simon & Schuster International Group.

Bender, E.A. (1995) *Mathematical Methods in Artificial Intelligence*, IEEE.

Bernold, T. and Hillenkamp, U. (eds) (1989) *Expert Systems in Production and Services II*, Elsevier Science Publishers.

Bestougeff, H. and Ligozat, G. (1991) *Logic Tools for Temporal Knowledge Representation*, Simon & Schuster International Group.

Bielawski, L. and Lewland, R. (1991) *Intelligent Systems Design*, John Wiley.
Bonissone, P.P., Henorion, M., Kanal, L.M. and Lemmer, J.F. (1991) *Uncertainty in Artificial Intelligence 6*, Elsevier Science Publishers.
Bramer, M.A. (ed.) (1990) *Practical Experience in Building Expert Systems*, John Wiley.
Browne, J. (ed.) (1989) *Knowledge Based Production Management Systems*, Elsevier Science Publishers.
Browne, J., Harhen, J. and Shivnan, J. (1988) *Production Management Systems*, Addison-Wesley Publishing Group.
Bryant, N. (1988) *Managing Expert Systems*, John Wiley.
Burman, R.M. (1991) *Computerised Management of Manufacture*, John Wiley.
Burton, A.M. and Shadbolt, N. D. (1987) *POP-11 Programming for Artificial Intelligence*, Addison-Wesley Publishing Group.
Campbell, J. and Cuena, J. (ed.) (1989) *Perspectives in Artificial Intelligence*, John Wiley.
Carling, A. (1990) *Introducing Neural Networks*, John Wiley.
Castillo, E. and Alvarez, E. (1991) *Expert Systems: Uncertainty and Learning*, Computational Mechanics Publications, Southampton.
Chen, C.H. (ed.) (1996) *Fuzzy Logic and Neural Network Handbook*, IEEE and McGraw-Hill Professional Publishing Group.
Corbett, J, Meleka, J., Pym, Ch. and Dooner, M. (1991) *Design for Manufacture*, Addison-Wesley Publishing Group.
David, J.-M., Krivine, J.-P. and Simmons, R. (eds) (1993) *Second Generation Expert Systems*, Springer-Verlag.
Diaper, D. (ed.) (1989) *Knowledge Elicitation*, John Wiley.
Diederich, J. (ed.) (1990) *Artificial Neural Networks: Concept Learning*, IEEE Computer Society Press.
Doumeingts, G., Browne, J. and Tomjlanovich, M. (eds) (1991) *Computer Applications in Production and Engineering*, Elsevier Science Publishers.
Duffin, P.H. (ed.) (1988) *Knowledge Based Systems*, John Wiley.
Eckmiller, R. (ed.) (1990) *Advanced Neural Computers*, Elsevier Science Publishers.
Eckmiller, R., Hartmann, G. and Hauske, G. (eds) (1990) *Parallel Processing in Neural Systems and Computers*, Elsevier Science Publishers.
Ellis, C. (ed.) (1989) *Expert Knowledge and Explanation*, John Wiley.
Farreny, H. (1989) *AI and Expertise*, John Wiley.
Filipic, B. (1988) *Prolog User's Handbook*, John Wiley.
Finch, F.E. and Kramer, M.A. (1987) Expert system methodology for process malfunction diagnosis, in *10th World Congress on Automatic Control*, vol. 6, IFAC, Munich, July, pp. 298–302.
Flach, P.A. and Meersman, R.A. (eds) (1989) *Future Directions in Artificial Intelligence*, Elsevier Science Publishers.
Ford, N. (1989) *PROLOG Programming*, John Wiley.
Friederich, S. and Gargano, M. (1989) *Expert Systems Design and Development Using VP-Expert*, John Wiley.
Futo, I. and Gergely, T. (1989) *Artificial Intelligence in Simulation*, John Wiley.
Gardin, J.C., Guillaume, O., Herman, P. *et al.* (1988) *Artificial Intelligence and Expert Systems*, John Wiley.
Gero, J.S. (ed.) (1988) *Artificial Intelligence in Engineering, Proceedings of the 3rd International Conference on Applications of AI*, Palo Alto, CA, USA, 8–11 August, Elsevier Science Publishers.
Gero, J.S. (ed.) (1988) *Artificial Intelligence in Engineering–Diagnosis and Learning*, Computational Mechanics Publications, Southampton.
Gero, J.S. (ed.) (1988) *Artificial Intelligence in Engineering–Design*, Computational Mechanics Publications, Southampton.

Gero, J.S. (ed.) (1988) *Artificial Intelligence in Engineering–Robotics and Processes*, Computational Mechanics Publications, Southampton.
Gero, J.S. (ed.) (1989) *Artificial Intelligence in Design*, Computational Mechanics Publications, Southampton.
Gero, J.S. and Stanton, R. (eds) (1988) *Artificial Intelligence Developments and Applications*, Elsevier Science Publishers.
Ginsberg, M. (1993) *Essentials of Artificial Intelligence*, Morgan Kaufmann Publishers.
Goodman, I.R., Nguyen, H.T. and Walker, E.A. (1991) *Conditional Inference and Logic for Intelligent Systems*, Elsevier Science Publishers.
Greenwell, M. (1988) *Knowledge Engineering for Expert Systems*, John Wiley.
Guida, G. and Tasso, C. (eds) (1989) *Topics in Expert System Design*, Elsevier Science Publishers.
Guidelines for the Documentation of Computer Software for Real-Time and Interactive Systems, The Institution of Electrical Engineers, 1990.
Gupta, U.G. (ed.) (1991) *Validation and Verification of Expert Systems*, IEEE Computer Society Press.
Hall, P.J. (1989) *How to Solve it in LISP*, John Wiley.
Harmon, P. and Sawyer, B. (1990) *Creating Expert Systems for Business and Industry*, John Wiley.
Harmon, P., Maus, R. and Morrissey, W. (1988) *Expert Systems*, John Wiley.
Hickman, F.R., Killin, J.L., Land, L., Mulhall, T., Porter, D. and Taylor, R.M. (1989) *Analysis for knowledge based Systems*, John Wiley.
Hollnagel, E. (1989) *The Reliability of Expert Systems*, John Wiley.
Houstis, E.N., Rice, J.R. and Vichnevetsky, R. (eds) (1990) *Intelligent Mathematical Software Systems*, Elsevier Science Publishers.
International Conference on Expert Planning Systems, The Institution of Electrical Engineers, 1990.
Ishida, H. (ed.)(1990) *InfoJapan'90*, Elsevier Science Publishers.
Jiang, J. and Doraiswami, R. Performance monitoring in expert control systems. AC10, vol. 6, 303–307.
Johnson, R.C. and Brown, C. (1988) *Cognizers*, John Wiley.
Kartalopoulos, S.V. (1996) *Understanding Neural-networks and Fuzzy Logic: Basic Concepts and Applications*, IEEE.
Klein, M. and Methlie, L. (1990) *Expert Systems*, Addison-Wesley Publishing Group.
Kline, P.J. and Dolins, S.B. (1989) *Designing Expert Systems*, John Wiley.
Kohonen, T., Mäkisara, K., Simula, O. and Kangas, J. (eds) (1991) *Artificial Neural Networks*, Elsevier Science Publishers.
Koschmann, T. (1990) *The Common LISP Companion*, John Wiley.
Kurki, M., Majanne, Y. and Ruokonen, T. (1991) Real-time expert system for power plant diagnosis, in *Third Sumposium on Export Systems Application to Power Systems*, Tokyo, April, pp. 160–164.
Lenzerini, M., Simi, M. and Nardi, D. (1990) *Inheritance Hierarchies in Knowledge Representation and Programming Languages*, John Wiley.
Liebowitz, J. (1991) *Institutionalizing Expert Systems*, Simon & Schuster International Group.
Lucas, P. and van der Gaag, L. (1991) *Principles of Expert Systems*, Addison-Wesley Publishing Group.
Milacic, V.R. and Majstorovic, V.D. (eds) (1989) *Intelligent Manufacturing System III*, Elsevier Science Publishers.
Morgan, N. (ed.) (1990) *Artificial Neural Networks: Electronic Implementations*, IEEE Computer Society Press.

Mumford, E. and MacDonald, B. (1989) *XSEL's Progress*, John Wiley.
Neapolitan, R.E. (1990) *Probabilistic Reasoning in Expert Systems*, John Wiley.
Nebendahl, D. (ed.) (1988) *Expert Systems*, John Wiley.
Odawara, G. (ed.) (1989) *CAD Systems Using AI Techniques*, Elsevier Science Publishers.
Parsaye, K. and Chignell, M. (1988) *Expert Systems for Experts*, John Wiley.
Parsaye, K., Chignell, M., Khoshafian, S. and Wong, H. (1989) *Intelligent Databases*, John Wiley.
Pau, L. (1987) Knowledge representation approaches in sensor fusion, in *10th World Congress on Automatic Control*, IFAC, Munich, July, vol. 6, pp. 319–323.
Payne, E.C. and McArthur, R. (1990) *Developing Expert Systems*, John Wiley.
Pedersen, K. (1989) *Expert Systems Programming*, John Wiley.
Pfau-Wagenbauer, M. and Brugger, M. (1991) Model and rule based intelligent alarm processing, in *Third Sumposium on Export Systems Application to Power Systems*, Tokyo, April, pp. 27–32.
Raeth, P.G. (1990) *Expert Systems: a Software Methodology for Modern Applications*, IEEE Computer Society Press.
Reeve, M. (ed.) (1989) *Parallel Processing and Artificial Intelligence*, John Wiley.
Ringland, G.A. and Duce, D.A. (ed.) (1988) *Approaches to Knowledge Representation*, John Wiley.
Rzevski, G. (ed.) (1989) *Artificial Intelligence in Manufacturing*, Computational Mechanics Publications, Southampton.
Rzevski, G. and Adey, R.A. (1991) *Applications of Artificial Intelligence in Engineering VI*, Computational Mechanics Publications, Southampton.
Salwa, K., Abd-El-Hafiz and Basili, V.R. (1995) *A knowledge based Approach to Program Understanding*, Kluwer Academic Publishers.
Sami, M. and Calzadilla-Daguerre, J. (eds) (1990) *Silicon Architectures for Neural Nets*, Elsevier Science Publishers.
Schildt, G.H. and Retti, J. (1991) *Dependability of Artificial Intelligence Systems (DAISY_91)*, Elsevier Science Publishers.
Shapiro, S.C. (ed.-in-chief) (1990) *Encyclopedia of Artificial Intelligence*, vols 1 and 2, Wiley-Interscience.
Sharda, R., Cheung, J.Y. and Cochran, W.J. (eds) (1991) *knowledge based Systems and Neural Networks: Techniques and Applications*, Elsevier Science Publishers.
Shi-quan Su (1987) An architecture for human-computer knowledge processing system, in *10th World Congress on Automatic Control*, IFAC, Munich, July, vol. 7, pp. 368–71.
Simi, M. and Nardi, D. (1990) *Inheritance Hierarchies in Knowledge Representation and Programming Languages*, John Wiley.
Sombe, L. (1990) *Reasoning under Incomplete Information in Artificial Intelligence*, John Wiley.
Soucek, B. (1988) *Neural and Massively-Parallel Computers*, John Wiley.
Soucek, B. (1989) *Neural and Concurrent Real-Time Systems*, John Wiley.
Sriram, D. (ed.) (1987) *Artificial Intelligence in Engineering – Tools and Techniques*, Computational Mechanics Publications, Southampton.
Sriram, D. and Adey, R.A. (eds) (1987) *Knowledge Based Expert Systems for Engineering – Classification, Education and Control*, John Wiley.
Sriram, D. and Adey, R.A. (eds) (1987) *Knowledge Based Expert Systems in Engineering – Planing and Design*, Computational Mechanics Publications, Southampton.
Steele, Jr, G.L. (1990) *Common LISP*, John Wiley.

Steels, L. (ed.) (1995) *The Biology and Technology of Intelligent Autonomous Agents*, Springer-Verlag.
Tatar, D.G. (1987) *A Programmer's Guide to Common LISP*, John Wiley.
Tello, E.R. (1989) *Object Oriented Programming for Artificial Intelligence*, Addison-Wesley Publishing Group.
Thayse, A. (ed.) (1991) *From Natural Language Processing to Logic for Expert Systems*, John Wiley.
Turban, E. and Watkins, P.R. (eds) (1988) *Applied Expert Systems*, Elsevier Science Publishers.
Vadera, S. (1989) *Practical Applications of Expert Systems*, John Wiley.
Vemuri, V. (1988) *Artificial Neural Networks: Theoretical Concepts*, IEEE Computer Society Press.
Wah, B.W. and Li Guo-Jie (1986) *Computers for Artificial Intelligence Applications*, IEEE Computer Society Press.
Wah, B.W. and Ramamoorthy, C.V. (eds) (1990) *Computers for Artificial Intelligence Processing*, John Wiley.
Walters, J.R. and Nielsen, N.R. (1988) *Crafting knowledge based Systems*, John Wiley.
Wang, D., Cheng, R. and Yang, Z. (1987) Computer aided scheduling system in integrated iron and steel works, in *10th World Congress on Automatic Control*, IFAC, Munich, July, vol. 7, pp. 378–83.
Wehenkel, L., van Sutsen, Th. and Ribbens-Pavella, M. (1987) Artificial intelligence applied to on-line transient stability assesment of electric power systems, in *10th World Congress on Automatic Control*, IFAC, Munich, July, vol. 6, pp. 308–13.
Wertz, H. (1988) *An Introduction to Programming in LISP*, John Wiley.
Whitby, B. (1988) *Artificial Intelligence*. John Wiley.
Widman, L.E. (ed.) (1989) *Artificial Intelligence, Simulation and Modeling*, John Wiley.
Winstanley, G. (1988) *Program Design for Knowledge Based Systems*, John Wiley.
Winstanley, G. (1991) *Artificial Intelligence in Engineering*, John Wiley.
Winston, P.H. (1992) *Artificial Intelligence*, Addison-Wesley Publishing Group.
Yen, J., Langari, R. and Zadeh, L.A. (eds) (1995) *Industrial Applications of Fuzzy Logic and Intelligent Systems*, IEEE.
Yoshikawa, H. and Gossard, D. (eds) (1989) *Intelligent CAD*, I, Elsevier Science Publishers.
Zahedi, F. (1992) *Intelligent Systems for Business*, Addison-Wesley Publishing Group.

Part Two

The Application of Intelligent knowledge based Systems in Electrical Power Engineering

3

Comments on contemporary power systems and other engineering systems

Jerzy S. Zielinski

3.1 MAIN CHARACTERISTICS OF ELECTRICAL POWER SYSTEMS

Electricity plays a very important role in contemporary society, which necessitates a reliable supply within the statutory parameters of electrical energy. To fulfil the above requirements for all consumers it is necessary to develop a power system over a very wide area and ensure continuous operation.

The most important properties of an electrical power system are as follows [1, 2]:

1. heterogeneity of technologies:
   (a) generation from hydro, thermal and nuclear power plants (wind and solar sources are considered as future forms of bulk electricity generation);
   (b) transmission and distribution of electrical energy with overhead lines and cables, with rated voltages exceeding 1 MV (AC or DC)
   (c) use of electrical energy by individual consumers and industry, where some require the energy with very little variation in parameters;
2. wide geographical area of power system operation;
3. practically simultaneous generation and use of the electrical energy (the lack of a method for electrical energy storage is of practical importance).

The above characteristics of a power system explain the need to use the best technological solutions in constructing the plant items, and the need for the application of computers in each stage of power system development, i.e. in research, design, construction and operation. It is necessary to remember though, that some computer applications in power systems need special attention, e.g. control in a bulk power system. Additionally, investment costs in power systems are high and proper forecasting and planning methods, as well as careful load-flow control, are required.

3.2 PROBLEMS IN ELECTRICAL POWER SYSTEM OPERATION

Power system operation depends on the state of the system, which necessitates proper methods of forecasting and planning. These methods have to be supplied with a wide set of data measured from the real power system in strictly defined intervals of time. This means that in power system operation, one of the problems is data acquisition. This measured data is the basis for all operational activities in power systems, which include the following operational problems:

1. **load and voltage control** linked with an economic dispatch and a security-constrained dispatch, which contains a number of operational tasks;
2. **fault diagnosis,** including alarm processing, protection, power system restoration and repair of damaged equipment;
3. **environmental protection** against power plant operation, and electromagnetic pollution from the electrical energy system;
4. **training** (especially dispatcher training) for selected groups of power system personnel;
5. **research and development** concerning all elements of the power system, the goal being maximum economic and efficient operation with minimal environmental pollution.

The majority of the aforementioned problems require to be solved on-line. For example, relay protection operates in the timescale of electromagnetic transients, and so needs a reaction time in the order of microseconds.

Successful solutions for the operational problems listed above need extensive support from modern, reliable computer systems.

3.3 COMPUTER APPLICATIONS IN POWER SYSTEMS

3.3.1 A Short history

According to Zielinski [3], computers have been used in power systems from the beginning of the computer age, and the following indicate the areas of application:

1. calculations and computer-aided design (CAD) (third-generation computers);
2. computer-aided manufacture (CAM) and control of power systems: both considered as control;
3. digital relay protection.

Computer applications in power systems can be characterized as follows. Hardware and software developments are applied as soon as possible, and the tendency is to interconnect all of the computer-aided functions (in a common system).

Developments in the late 1960s included automatic generation control (AGC), the load management control system (LMCS), direct load control (DLC), the energy control system (ECS) and others. Gradually the development of the above systems led to supervisory control and data acquisition (SCADA) systems, and subsequently to energy management systems (EMSs).

EMSs are the most-developed computer systems applied in power systems, and the references present a number of these systems operated by different companies. As good examples, consider the EMS developed by the Houston Lighting & Power Co. [4] or KEPCO's national control centre with its advanced EMS [5]. This second reference presents very interesting data regarding EMSs: architecture, functions, major features, system performance (ultimate system sizes, CPU loading requirements, display response time requirements, actual CPU loading, actual display response time) and actual operating experience.

3.3.2 Some remarks on hardware and software

The list of functions realized by a contemporary EMS is very long (e.g. Kim *et al.* [5]), and contains functions of varying importance. The following three categories of functions are considered in Barrie *et al.* [2].

1. The highest-priority functions are involved in the protection systems and in the monitoring of the power system. These functions need computer system availability to be no less than 99.94%, which means a break time of no longer than 5 hours per year.
2. The high-priority functions are associated with the program of security assessment. These functions need computer system availability to be no less than 99.9%, which is equivalent to 10 hours per year break time.
3. The medium-priority functions, for forecasting and planning, need computer system availability to be no less than 99.6%. This is equivalent to 30 hours per year break time.

Additional information on computer system loading is found in Horton and Grass [6]. For example, state estimation for a 500-substation system takes 5–10 s repeated in 5–30 min intervals, and optimal load flow for the same system takes 70–100 s repeated in 60–90 min intervals (on a 1 MIPS (million instructions per second) processor). A typical EMS would exhibit the following processor speeds [6]: 0.7 MIPS in 1980, and 12.47 MIPS in 1985.

As a result of the large computer system loading and the high availability required in an EMS, it is necessary for redundancy to be built into the computer systems architecture, except in the case of displays and links with remote terminals [6]. Most of these systems are distributed computer systems which have a number of advantages [2] over the systems with a cluster architecture.

It is very interesting to note that the expected life of an EMS is considered to be about 15 years [7], or 10–15 years [6].

The efficiency of an EMS also depends on the programming language used. Furthermore, it is very important to remember that an EMS needs large databases with very high availability.

3.3.3 IKBSs

The last decade of computer applications in power systems has witnessed a significant number of IKBS developments. The first IKBS in this domain was developed to monitor vibrations in order to diagnose faults in large pumps in a nuclear power station, and has been in operation since 1980. The accident at the Three Mile Island nuclear power station in 1979 highlighted the human limitations in the operation of such complex systems, and as a result started the age of knowledge based system applications in power systems.

Since then, the field of expert system applications in electrical power engineering has grown at a significant rate. Data concerning IKBSs in power systems in Europe is collected in Germond and Niebur [8]. According to this paper, the percentage spread of the 40 projects described is:

1. alarm reduction and system diagnosis, 27.5%
2. environments for operational aids, 12.5%
3. steady-state security and dynamic security, 15%
4. management, scheduling and planning, 15%
5. component diagnosis, 2.5%
6. remedial controls, 10%
7. restoration, 10%
8. substation monitoring and control, 7.5%.

The distribution of the means of implementing the above IKBSs among the various software techniques is:

1. object-oriented tools, 5 IKBSs
2. AI toolkits, 12 IKBSs
3. AI languages, 11 IKBSs (Prolog, 8; LISP, 2; OPS83, 1)
4. algorithmic languages, 12 IKBSs (Pascal, 6; Fortran, 3; other, 3).

The hardware used for the above IKBS implementations is distributed as follows:

1. general purpose workstations, 40%
2. PCs, 25%
3. LISP-based workstations, 15%
4. mainframes, 5%
5. SCADA computers, 15%.

Another interesting perspective, which is based on responses received from 24 organizations all over the world, is presented in Liu and Van Son [9] and will be briefly reported below. The motivating issues behind IKBS application, as indicated by respondents, was said to be:

1. to prevent human error, 4
2. to maintain system knowledge, 3
3. to standardize actions, 5
4. to reduce operator burden, 11
5. to test IKBS techniques, 17
6. to improve inefficient analytical programs, 5.

This same survey found that IKBSs were applied for:

1. fault diagnosis, 11
2. restoration, 9
3. alarm processing, 6
4. operation assistant, 3
5. remedial control and security assessment, 5.

The following were said to be involved in the development process of a knowledge base.

1. Developers: experts (outside and/or own personnel), operators, end-users, database managers, consultants.
2. Knowledge acquisition techniques: interview, case study, fundamentals of protection, rapid prototyping and frequent user evaluation, verbal protocol analysis method, testing with analytical tools, published work.
3. Workforce requirements: based on available funding, 5 person–years, 18 person–months (lab-prototype) and 55–person-months (prototype), 5–year period, 3–year period, 73–person–months.

4. Phasing-in: generally according to typical development procedure.

The results of the IKBS performance evaluation follow.

1. Desirable/undesirable responses (%): from 75/25 to 100/0.
2. Speed: from 2 s to 2 min, or a few times faster than other methods.

The user evaluation of the IKBS tools was as follows.

1. Acceptance by the user: satisfied, only one respondent considered their IKBS development tool difficult to use.
2. Evaluation by organization: positive.

These applications adopted the following knowledge engineering features.

1. Rule chaining: forward, 10; backward, 2; mixed, 6.
2. Database: relational.

Finally, the response of management and justification were as follows.

1. Management view: positive.
2. Cost: hard to quantify, $50 000, $450 000, benefit–cost analysis of 4/1, 7 person–years.

Another point of view on some of the problems can also be found in Van Son and Sforna [10].

The details of particular IKBS applications in power systems will be developed in Chapters 4–9.

3.4 SOME REMARKS ON THE RELATIONSHIPS BETWEEN PHENOMENA IN POWER SYSTEMS AND OTHER FIELDS OF ENGINEERING

It is appropriate to mention some of the relationships which exist between phenomena in electrical power systems and other fields of engineering with similar types of equations describing their processes. As is well known, electromagnetic wave motion is described by hyperbolic quasilinear partial differential equations, and these are the basis of the transient analysis of transmission lines, transformers, rotating machines, coils etc. It is, however, possible to use simplified ordinary differential equations in a number of practical situations, e.g. steady-state analysis. A similar approach is also adopted in other fields of engineering, such as in mechanics and hydrodynamics. Thus, the same types of equations are used in a number of fields of engineering, and only the parameters in these equations are different because of their physical origin.

Examples of processes of a similar nature in different fields of physics and engineering can be found in the references:

1. the analogy between quantities in scalar potential fields [2,11];
2. the analogy between mechanical and electromagnetic elements [12,13];
3. the equivalence of variables and parameters in physical fields, and characteristics of partial differential equations [13,14];
4. the analogy between mechanical and electromechanical elements, between electrical and hydrodynamic fields, between long transmission lines and pipelines, and between heating systems and electrical circuits [13].

It is also important to remember that electrotechnics was developed as a part of mechanics, and, by tracing this history, methods and tools for solving practical problems can be found (e.g. the momentum principle).

3.5 CONCLUSIONS

Consideration was given in Chapter 2 to showing the importance of IKBS applications in industry and to presenting their functions (section 2.2) and the general trends exhibited in their development. Similarly, sections 3.2 and 3.3 presented important functions and trends in IKBSs applied to power systems. Furthermore, it was identified that there are a number of similarities between power systems and other engineering disciplines. This means that experiences in the development of IKBSs for power systems, e.g. for alarm processing or real-time control, may be applicable to other fields of engineering. This is especially true when the relationships between power systems and the other fields of engineering considered in section 3.4 are taken into account.

REFERENCES

1. Zielinski, J.S. (1991) Some remarks on artificial intelligence and computers, in *Proceedings of Seminar on 'Expert Systems and Integrated Case-Tools'*, Lodz, September, pp. 30–38.
2. Barrie, D., Hill, D.S. and Yuen, A. (1989) Computer configuration for Ontario Hydro's new management system. *IEEE Transactions on Power Systems*, **4**(3), 927–34.
3. Zielinski, J.S. (1987) Some remarks on digital computer applications in power systems, in *Seventh International Conference*, Ostrava, CSRS, 5–8 April, 371–75.
4. Trefny, P.J., Gross, D.P., Wollenberg, B.F. and O'Hehir, J.D. (1988) Integration of an operations information system into energy management and control system. *IEEE Transactions on Power Systems*, **3**(1), 262–66.
5. Kim, Y.H., Fukushima, N. and Dy Liacco, T.E. (1990) KEPCO's national control center with an advanced energy management system. *IEEE Transactions on Power Systems*, **5**(4), 1084–90.
6. Horton, J.S. and Grass, D.P. (1987) Computer configurations. *Proceedings of the IEEE*, **75**(12), 1659–69.

7. Hanson, K. (1988) Status report on control centre maintenance practices. *Electra*, No. 117 (March), 46–54.
8. Germond, A.J. and Niebur, D. (1992) Survey of knowledge based systems in power systems: Europe. *Proceedings of the IEEE*, **80**(5), 732–44.
9. Liu, C.-C. and Van Son, P. (1993) Expert system development experience and user requirements. *Electra*, No. 146 (February), 29–67.
10. Van Son, P. and Sforna, M. (1993) Exploring user requirements of expert systems in power system operation and control. *Electra* No. 146 (February), 69–84.
11. Zhang, Z.Z., Hope, G.S. and Malik, O.P. (1989) Expert systems in electric power systems – a bibliographical survey. *IEEE Transactions on Power Systems*, **4**(4), 1355–62.
12. Binns, K.J. and Lawrensen, F.J. (1963) *Analysis and Computation of Electric and Magnetic Field Problems*, Pergamon Press, Oxford.
13. Zielinski, J.S. (1980) *Analogue and Digital Modelling*, textbook, Technical University of Lodz, Lodz, Poland (in Polish).
14. Huskey, H.D. and Korn, G.A. (1962) *Computer Handbook*, McGraw-Hill Book Co.

4

Forecasting and planning in power systems

U. George Knight

4.1 INTRODUCTION

In order to give a balanced presentation, while noting that this book is addressed to a mixed audience, this chapter commences with some background on planning and forecasting before going on to give examples of the application of expert systems. The contents of this chapter consequently cover the following: planning tasks in the extension of power systems; forecasting tasks in the extension of power systems; the evolution of techniques for system planning and forecasting; the characteristics of system planning and forecasting tasks in relation to expert systems; areas of research and development; descriptions of some proposed and actual applications of expert systems in generation expansion planning, in transmission network planning, in distribution network planning, and in power systems analysis; demand forecasting; trends and conclusions.

4.2 PLANNING TASKS IN THE EXTENSION OF POWER SYSTEMS

As far as the author is aware, there is no uniformly accepted definition of system planning. The tasks covered vary between utilities, principally when considering the design of the transmission system and the time span covered. However, the term is usually also taken to include reinforcement to the generation as well as the transmission and distribution systems.

The overall planning process can fundamentally be considered in two phases [1] as follows.

1. The determination of parameters for plant which meet the system requirements **on average** in the best way (plant standardization

studies); the resulting standards will typically be retained for 10 years or more.
2. Design of specific reinforcements using standard plant ratings; this will be a continuous process with the reinforcements timed mainly to meet increases in demand, connections for new generating plant, and corresponding increases through the body of the system.

The specific tasks involved will be as listed below (the way in which these are approached and the importance attached to them will depend significantly on the organizational structure of the industry):

1. determination of the capacity of new generating plant to be installed year by year;
2. determination of the types, unit sizes and locations of this new generating plant;
3. decisions regarding existing generating plant to be scrapped (or less frequently undergo major refurbishment);
4. assessment of the type of transmission and main structure of the transmission system, including transforming points to subtransmission and distribution networks and the timing of extensions;
5. setting of the main parameters of the transmission plant – voltages, throughput and fault ratings, configuration of substations;
6. determination of transmission network extensions;
7. setting of the main parameters of the distribution networks;
8. determination of distribution network extensions.

4.3 FORECASTING TASKS IN THE EXTENSION OF POWER SYSTEMS

Studies for time spans of up to 20 years require a wide range of forecasts. For each of the selected years of this period these will include the following.

1. Demand – estimates of the peak and minimum demands, the demand cycle or a demand duration curve, measures of the variability of these estimates and the geographical distribution of demand including new supply points. Energy and active power estimates will be required for the whole utility and active power and reactive power (or power factor) estimates for local areas and individual supply points. Some analytical studies may require a deeper knowledge of the structure of the demand, e.g. proportions of heating, motive power etc. and the responses of the demand to variations in voltage and frequency.
2. Generation – potential new sites for generation and capital costs, construction times; candidate plant for scrapping; estimates of peak and average availabilities (perhaps maintenance periods, forced and

planned outage rates), operating efficiencies, technical parameters.
3. Fuel – estimates of availabilities and prices of fuels.
4. Transport – tonnage, routes and costs.
5. Transmission – estimates of capital costs for new plant; failure rates and maintenance periods for new and existing plant, technical parameters.

It will be seen that forecasting embraces a wide range of topics. The introduction of new technologies and sometimes major changes in ratings will require in depth studies and probably a prototype stage. Some years ago the author described forecasting as the core problem of planning but its importance seems to be decreasing a little with the organizational and technical changes of recent years.

4.4 THE EVOLUTION OF TECHNIQUES FOR SYSTEM PLANNING AND FORECASTING

System planning and forecasting engineers were amongst the earliest users of analytical and optimization techniques in the power systems field. A use of probability analysis to relate generation margin to reliability of supply was first proposed in 1947. A paper describing the application of digital computers to power system analysis was published in 1952 [2]. Improved formulation of power system analysis programs has continued to the present day, to the point at which load flows can be computed for networks with thousands of nodes in minutes, and transient behaviour can be estimated in real time. Such a powerful analytical background is important when considering expert system applications since it means substantial algorithmic support can be provided to an expert system, for instance sensitivity factors relating flows in circuits or voltages at substations to changes in system configuration or demand levels, etc. Nevertheless there are outstanding problems: the requirement for transient stability techniques sufficiently fast and accurate to allow a rigorous search to be made to identify critical fault locations and clearance times; the requirement for voltage and dynamic stability assessments extending over protracted periods of time on very large systems; the availability of unit commitment techniques which provide the times, within say 15 min, at which units should be synchronized or desynchronized and include all the relevant plant and system constraints (although this is an optimization rather than analytical problem).

Optimization techniques have been studied since the 1950s (e.g. Massé [3]). Linear programming is the most widely used; important applications include the selection of type and capacity of generating plant and sometimes location over a period of time. Dynamic programming has also been used and problem-specific nonlinear programming models have been developed.

The synthesis of 'optimum' configurations is an intractable problem; the constraints needed to ensure system security lead to a combinatorial-type formulation. Nearly all the models proposed have been heuristic in total or part, sometimes employing linear or dynamic programming techniques. In practice transmission reinforcement schemes are devised/proposed by the planner, often using an interactive suite of system analysis programs to test alternatives and prove the viability of the system.

Prior to the early 1970s, demand forecasting for the planning timescales was achieved mainly by trend estimates. Results were critically dependent on the growth curves assumed; those fitted to demands recorded over the past 40 years could be indistinguishable to the eye and yet when extrapolated for the next 10 years show increases varying between 1½ times and twice the last actual value. Forecasting models based on much wider considerations are now used, for instance [4]:

1. trends mixed with economic modelling;
2. forecasts of the growth of the gross domestic product (GDP) and of the ratio between GDP and use of electricity;
3. splitting of total GDP growth into growth per capita of population;
4. estimation of total primary and secondary (delivered to final consumer) energy requirements and dependence on GDP;
5. estimated changes in the efficiency of use by the final consumers of electrical energy;
6. estimation of market shares taken by competing fuels (this may be done for the total market or for market sectors);
7. comprehensive models of the fuel sector and of the whole economy.

This evolution in demand forecasting techniques is important to the prospective use of expert systems in this area, since it means that the knowledge base is diffuse with decisions involving considerable judgement plus harmonization of many factors and views. Longer-term forecasts now often include the concept of 'scenarios', e.g. high, medium and low growth estimates at a putative year (e.g. Davies [4]).

4.5 THE CHARACTERISTICS OF SYSTEM PLANNING AND FORECASTING TASKS IN RELATION TO IKBSs

Prior to describing some of the applications of IKBSs in the remainder of this chapter, it is worthwhile considering the characteristics of these tasks in relation to the problem solving characteristics of expert systems and of the older, more conventional techniques. This will help to clarify the level of interest of the supply industry and IKBS practitioners in the various fields. Comments on this and brief descriptions of many expert system proposals are included in Knight [5].

4.5.1 Algorithmic, human and expert system capabilities

In broad terms, the strong and weak points of algorithmic, human and IKBS approaches to problem solving will be as follows.

(a) Algorithmic

Strong points:

1. number crunching;
2. auditable results, e.g. provable optimum solutions for the formulation used (an important qualification);
3. very strong analytical background.

Weak points:

1. must avoid formulations leading to 'combinatorial explosions' (these often result from the obvious approach);
2. difficulty in handling 'fuzzy' problems and data;
3. difficult to incorporate problem aspects which cannot be quantified.

(b) Human

Strong points:

1. can identify patterns and connections, can distinguish the 'wood from the trees' and hence can isolate profitable lines of attack;
2. can make selective use of numerical analysis;
3. can correlate data from a very wide base of knowledge and experience;
4. is intuitive.

Weak points:

1. very limited arithmetic capability;
2. may be difficult to validate conclusions;
3. reactions and conclusions may not be repeatable;
4. may be difficult to extract line of reasoning.

(c) IKBSs

Strong points:

1. can emulate some aspects of the human approach, in particular to limit areas of search;

2. can handle fuzzy data and logic;
3. can be linked with algorithmic formulations, e.g. to provide sensitivity solutions;
4. can indicate line of reasoning.

Weak points:

1. may be difficult to validate conclusions;
2. difficult to incorporate 'common sense';
3. difficult to incorporate complicated, highly interactive strategies;
4. knowledge relating to the problem must be obtained from experts in the field (this is time consuming and possibly contradictory).

4.5.2 The characteristics of planning and forecasting tasks in relation to IKBSs

The complementary question to those of the previous sections is what are the characteristics of the main tasks in the domains of power system planning and forecasting. These can be split broadly into synthesis – that is postulating the system design which should meet requirements, e.g. demand and constraints met at minimum cost; and analysis – that is determining whether the potential design does satisfy the constraints. In mathematical terms, structural optimization problems often result in integer combinatorial formulations, time consuming and difficult, if not impossible to solve formally. On the other hand an experienced engineer or operator will be good at postulating new and adjusting existing configurations. Intelligent knowledge based systems can be devised to support this ability. Conversely, algorithmic techniques are in general well suited to optimization or analysis of a system whose structure has been defined.

The task of forecasting would seem to be partly algorithmic, although logical processes may well be involved in selecting appropriate parameters for the algorithmic processes. There will be substantial elements of judgement.

A basic appreciation therefore of the potential of IKBSs in planning and forecasting from a technical point of view is as follows.

1. Generation capacity – this is essentially algorithmic once security standards have been set. The two phases may be combined, for example in utilities in which cost of loss of supply is quantified and included in a system cost objective function. Scrapping of old or inefficient plant must be considered.
2. Generation mix – this, the problem of determining what types of generating plant should be built, is also algorithmic in nature.
3. Generation location – specific sites must be selected for new generating plant, including extension or replacement at existing sites. It is

possible in exploratory, longer-term studies for the number of possibilities to exceed considerably the number required and the problem can become combinatorial. It is also, of course, integer.

4. Transmission network extensions – the extensions to the transmission network to connect new generation and reinforce capacity to the lower voltage networks must be planned. The impact of plant retirements must be considered. The studies may be definitive, covering say the next 3–10 years, or long term and indicative extending over 20 years and longer. These will include such major developments as the introduction of a new transmission voltage or direct current links. In mathematical and logical terms, this is the most difficult task, combinatorial, highly interactive and often involving some form of pattern recognition.
5. Composite model – numerous models and suites of programs combining two or more of these tasks have been developed and some are commercially available (e.g. references 6 and 7). The most common combination is to study generation capacity and mix.
6. Subtransmission and distribution network extensions – the task is to determine all the extensions needed to the networks between the transmission system and the end consumers. Typically three or four voltage levels will be involved. Although conceptually the same as the transmission case, in mathematical and logical terms the problem is less complex because distribution networks are usually radial and the security criteria less demanding. Although there are many more nodes on distribution than transmission systems, distribution networks are usually supplied from only one or two transmission substations and there is virtually no interaction between changes implemented within the distribution network supplied from one transmission point and those from other transmission points.
7. Transmission and distribution plant standardization – approaches to this range from manual in which experienced engineers design and cost, in some detail, networks assuming plant of different ratings, perhaps voltages etc., to a mathematical model in which, starting with demand, area of supply and security criteria, equipment requirements are estimated for different ratings, voltage levels etc. This is basically a constrained optimization formulation, any questions of structure having been included in equations used to estimate the equipment requirements.
8. System analysis – many models are available to estimate the steady state and dynamic performance of power systems at various levels of detail. The problems lie in determining the critical conditions for analysis (e.g. demand level, generation distribution and fault location in a transient stability study), sifting and digesting the outputs and for some types of studies, determining system and plant data.

Although the analysis is obviously algorithmic in nature, the preparation of data, choice of cases for analysis and sorting of output are areas in which support to the planner can be very useful.

A summary of the foregoing relating the adequacy of conventional computer aids and the judged scope for IKBSs in the various tasks is given in Table 4.1.

4.5.3 Frequency of study

So far the emphasis has been on technical aspects. It seems to the author that another important feature is the frequency at which a particular investigation has to be made. For instance, is it worthwhile developing an IKBS for a task which comes up infrequently, e.g. major change in the structure of the transmission network? The answer may well depend on the role of the organization considering the application of IKBS. A utility, faced with such a change at 10-year intervals or more, might feel it would be better to use its resources for the specific problem rather than to develop an IKBS. A consulting company, faced with the possibility of having to investigate this problem much more frequently for various companies, could well take the opposite view. One may also find that the development of a model which essentially mechanizes part of the outline design or standardization process will allow many more alternatives to be considered. This is demonstrated in Knight [1] for subtransmission and distribution standardization using a heuristic approach, and in Galiana *et al.* [8] and Loud *et al.* [9] for outline transmission studies using an IKBS approach. As another example, a utility with responsibilities for distribution might have to authorize every year hundreds of reinforcements and extensions at the lower voltage levels with staff involved across its whole area and could feel that IKBS support to this staff would be justified.

4.5.4 Retention of expertise

Another factor which may influence a decision to develop IKBSs is the preservation of staff expertise. Planning and forecasting are both areas in which judgement and experience are important to harmonize quantitative model solutions with qualitative environmental, social etc. issues. Planning activity may also be cyclical, following the economy, and particularly when low, staff may move to other work. IKBSs provide a way to record and access knowledge which cannot be included in conventional algorithmic models and also to conserve knowledge which might be lost as a result of staff movements. In fact one definition of an IKBS is 'a means of recording and recalling at will human competence

Table 4.1 Assessment of solution techniques for various tasks

Task: determination of	*Problem type*	*Comment on conventional computer-aided techniques*	*Apparent scope for application of IKBS*
Generation capacity	Algorithmic, may include optimization	Adequate	Limited
Generation mix	Algorithmic, constrained optimization	Adequate	Possibly
Generation location	Combinatorial, integer	Usually done by series of manually directed studies, computer used for costing	Yes
Transmission extensions	Combinatorial, integer, logical, elements of pattern recognition	Usually done by series of manual studies, computer used for analysis and costing	Yes
Composite model	Combines elements of capacity, mix, and possibly location as appropriate	Reasonable for capacity and mix. Strong manual element in location and transmission extensions	Yes. Applications could also include organization of data and sequence of studies
Subtransmission and distribution extensions	Combinatorial, integer, logical, elements of pattern recognition	Often done by series of manual studies, computer used for analysis	Yes
Equipment standards	Algorithmic, bounded variable optimization	Often done by series of manual studies, with computer used for analysis and costing. Mathematical models have been developed	Not obvious (but see section 4.5.3)
System analysis	Algorithmic	Models exist for all types of steady-state and dynamic behaviour. Problems in selecting cases for study, analysing output, collecting and preparing data	Yes, in organizing data and studies, and in analysing outputs
Forecasting demand (long term)	Combination of various approaches including statistical	Data and information needs may be very wide ranging	Difficult problem. May not be suitable for an IKBS approach

in a particular specialist, usually narrow, field (often called domain). As such it can serve as an expert in that field'.

4.6 AREAS OF RESEARCH AND DEVELOPMENT

Research into power system applications of IKBSs started in the early 1980s (although these potential applications tended to remain in the research or development areas much longer than in some fields). Since then, proposals and projects have come forward at an increasing rate. These are mostly concentrated in the following areas.

1. System operation and control:
   (a) alarm analysis
   (b) fault location and analysis
   (c) system restoration
   (d) voltage and reactive power (VAR) control
   (e) short-term demand forecasting
   (f) scheduling
   (g) remedial switching.
2. System planning:
   (a) distribution network planning
   (b) VAR planning
   (c) transmission network design
   (d) generation expansion planning.
3. Network analysis:
   (a) data management
   (b) support in load flow studies
   (c) support in stability assessments.

The order in which the applications are shown in each list indicates the judged interest, based on number of papers published. Overall, proposals in the operational and control areas have well exceeded those in system planning and network analysis.

The distribution of development effort broadly supports the conclusions reached earlier concerning productive lines of work. Other factors which will influence the distribution will be as follows.

1. The severity of the problem as perceived by the end-user planner or operator, and hence the support given to the proposed research: it seems that operators have generally seen greater scope for the application of IKBSs than planners. For instance, alarm analysis and fault location are important tasks in operation, combining elements of pattern recognition and logical analysis, and requiring rapid solutions. There are no counterparts in planning.
2. Shorter solution times are often available from IKBSs compared to algorithmic methods.

Descriptions of applications in generation expansion planning, transmission network design and distribution network design have been included in the following sections. The descriptions provided are taken from published work, include references to the source material, and as far as possible will include objective, brief, descriptions of the knowledge base and examples of rules, output, implementation and status. It should be noted that inclusion of a particular IKBS does not imply its endorsement *vis-à-vis* others with a similar purpose which have not been included.

4.7 DESCRIPTIONS OF SOME PROPOSED AND ACTUAL APPLICATIONS OF IKBSs IN GENERATION EXPANSION PLANNING

There will be several components common to any technique, including IKBSs, for generation expansion planning:

1. An algorithm will be used to estimate the operating costs (fuel) for each system state (capacity and mix of plant) over each specified time interval (calculated yearly or for selected years over the total time span). This is often called a production costing algorithm and many models have been devised. These range from methods in which the demand distribution over the specified time interval (Figure 4.1) is convoluted with a generation cost distribution (Figure 4.2) to detailed models in which an hourly or half-hourly generation schedule and dispatch is computed and costed over representative days. This is usually the most computationally intensive part of the whole generation expansion model.

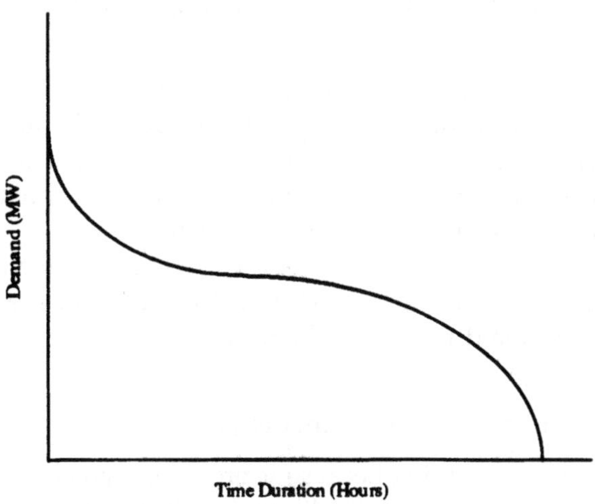

Figure 4.1 Demand duration curve.

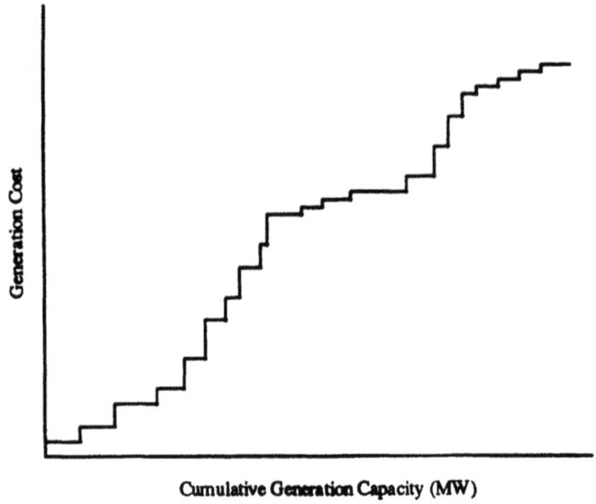

Figure 4.2 Cumulative generation cost curve.

2. Algorithms will also be used to estimate:
   (a) staff and maintenance costs (e.g. annually);
   (b) capital charges (e.g. annually);
   (c) the cost of energy or power not supplied (e.g. annually) because of breakdown of generating plant or higher than expected demands;
   (d) the present worth of the summated costs 1 and 2(a)–(c) over the total time span.
3. There will be a procedure to produce possible generation programmes for the selected years of the total time span. This may be done manually by planning engineers or by some optimization process, e.g. dynamic programming, linear programming, nonlinear programming or decision-tree analysis.

It should be noted that the planning engineer is usually interested in obtaining a number of solutions, hopefully including the minimum cost one, which satisfy the prescribed constraints and which will give him or her a broader base on which to make a decision incorporating judgemental factors not included in the formulation.

Two proposals will be described below; each uses some of the techniques outlined above combined with an IKBS to limit the number of possibilities considered (the **area of search**).

4.7.1 A decision tree and IKBS model [10]

This paper is introduced first because it provides a good background in the financial and costing aspects although only limited use is made of IKBSs.

(a) Objective

The objective of the model is to produce a number of possible generation expansion programmes, including the one with minimum total present worth, for each specified scenario (demand levels, fuel costs etc.). The output will include:

1. the optimum and near-optimum generation programmes, i.e. the capacities and types of generating plant to be installed in each year of the time span;
2. the total present worth of each of these programmes; one would assume that this total can be broken down into the year by year figures and their separate components.

(b) Data

The data requirements are as follows.

1. Financial: interest and repayment charges on capital costs, insurance, constraints on financing etc.
2. Technical:
   (a) annual demands over total time span (peak values and load factors);
   (b) existing generation (units with location, type, size, fuel and maintenance costs, fixed outage rates, limitations on operation);
   (c) candidate-generating units (units with location, proposed commissioning year, type, size, fuel and maintenance costs, immature and mature outage rates, limitations on operation);
   (d) system constraints (reliability, environmental).
3. Study details: scenarios (e.g. energy forecast (classified as high, medium or low), fuel supply), possible decision-tree paths.

(c) Model

The possible generation states are determined in the form of a decision tree and a number of rules applied to prune the branching of this tree so as to limit the computation to exploring reasonable paths of development. The rules suggested are as follows.

1. Each path is reviewed by a planning engineer and those considered not reasonable are deleted. An example of such a path would be one including too high a proportion of a particular type of generation.
2. Proposed decisions which either lead to system states not satisfying technical constraints (e.g. reliability) or which themselves contravene other constraints, for instance financial or environmental, are deleted.
3. Paths which lead to states whose present worth and component costs

are considered too high, by comparison with already determined paths, are deleted.

4. Only one of the paths which converge on an identical state needs to be retained.

A probabilistic production costing procedure is used. In the first stage the expected demand is reduced to simulate the generation from nonthermal sources. In the second stage a probabilistic convolution technique is used to estimate expected generation from each unit, the loss of load probability and the expected unserved energy.

(d) Implementation

A series of possible study environments are selected (for instance demand level, oil supply level, operation of heavy-oil-generating units) and the probability of each of these estimated. For each of these study environments an IKBS is used to evaluate the acceptability of the various decision paths. The acceptable paths are then listed in order of ascending present worth for each of the study environments. Finally, a ranking order covering all the study environments can be determined by weighting the present worth for each study environment by the probability of that environment and summating.

The IKBS rules were written in PASCAL as (if-then) statements to run on a PC.

(e) An example

Results are given from a study on the Egyptian systems [10]. This was for a planning period of 10 years (1991–2000), the initial system capacity being 8800 MW in 115 units comprising 22 hydro units (22–175 MW), 38 heavy oil units (12.5–300 MW) and 56 natural-gas units (20–49 MW). It was assumed that the extension plant programme would be selected from the 1000 MW nuclear, 300 MW and 100 MW combustion turbine units.

Twelve study environments were selected covering three demand levels at the year 2000: high (8185 MW), medium (7837 MW) and low (7503 MW), normal and reduced oil supply and normal and derated operation of heavy-oil units. The five planning programmes with minimum total present worths for each of the 12 study environments are retained for the calculation of the overall optima.

It is common experience that the cost variations of solutions around the optimum in most engineering systems are small and this is again demonstrated here. The cost variations across the five solutions of each of the study environments are only some 1–1½%, nevertheless representing large sums of money.

4.7.2 A combined dynamic programming, IKBSs and fuzzy sets approach

This approach [11–13] combines three significant features: the use of dynamic programming to track an optimal expansion programme; a rule-based IKBS to incorporate engineering judgement, experience and opinions to reduce the area of search (e.g. number of alternatives examined); fuzzy set theory to integrate qualitative (sometimes conflicting) requirements. As an illustration of these, a utility or country may decide to limit its reliance on a certain fuel, say imported oil or indigenous coal, for strategic and security reasons. There may be preference for certain suppliers of plant.

A further feature of the approach is that rules can be relaxed, for instance a loss of load probability in one year slightly higher than the maximum specified could be accepted in order to prevent a promising final-year solution being eliminated.

(a) Objective

Practical generation expansion programmes result from extensive analysis of selected alternatives and discussion between experienced power engineers. Frequently one of the infeeds to this process is the result from some form of heuristic or mathematical optimization program. Most models will only include the factors which the planning engineers feel they can quantify, e.g. costs, security criteria, but not the qualitative judgements such as good long-term potential, ease of site acquisition and planning permissions etc. The authors of this group of papers set out to demonstrate:

1. a dynamic programming formulation in which the dimensionality problems are kept within feasible bounds;
2. an integration of expert technical knowledge and judgements with a rigorous mathematical optimization method;
3. an ability to formulate and modify earlier rules derived from the planning engineers' knowledge and experience;
4. an ability to include qualitative judgements in a logically consistent manner, using a fuzzy-logic approach;
5. computational efficiency, through the use of Prolog, for the logic processing and bounded areas of search in the dynamic programming optimization;
6. flexibility in the solution, such as the ability temporarily to relax some constraints.

(b) Methodology

The methodology will be outlined in four parts: structure, dynamic programming optimization, role of the IKBS, and the role of fuzzy logic.

Structure

The basic data is organized into project frames and state frames. A project frame (PF) contains numerical and non-numerical information on each generating-plant option, for instance: type of plant (nuclear, coal, hydro etc.), location, first possible commissioning date, power rating, range of reactive power outputs, reactances, operating and capital costs, availability and forced outage rate, restrictions on stages of development and necessary relationships with other plant options (e.g. a nuclear unit must be accompanied by a gas turbine unit at the same site).

State frames (SFs) summarize uniquely each state of the system developed during the dynamic programming process, starting with the initial state at the base year. Hence state frame SF(*mn*) contains the data for time stage *m* and a unique combination denoted *n* of generation projects. It will also include some information common to all frames at this stage, such as the load forecast, and some specific to that state such as its costs which are dependent on the particular mix of plant.

Rules are classified as **window rules** or **transition control** rules. The former are those which relate to the state of the system, for instance the minimum plant capacities required in the states at each stage. The latter are those which define acceptable changes between stages (and also, hopefully, limit the number of these).

Dynamic programming process

The forward-looking dynamic programming process is used [14, 15]. In this, and in accordance with Bellman's principle of optimality, if $C_k(S_i)$ is the cost of state i in stage k, C^k_{k-1} the cost of the change from state j in stage $(k-1)$ to state i in stage k, and A_{k-1} the accumulated cost of being in stage j at the end of stage $(k-1)$, the recurrence equation defining the sequence of optimum changes is

$$A_k(S_i) = \min_j \left[A_{k-1}(S_j) + C_k(S_i) + C_{k-1}(S^i_j)\right] \tag{4.1}$$

The authors make the point that a better overall solution may be obtained if some constraint, whose action at one or two earlier time stages would prevent that final solution being obtained, were temporarily relaxed. An example of this might be that generation from the development of a highly desirable site could only be available one year later than required to maintain the specified planning margin, giving the options of either distorting the optimum generation expansion programme or accepting an increased probability of loss of load for one year. The technique proposed is to retain at the appropriate stages not only the single optimum state as defined by equation 4.1 but also the one (or more) states which include the desirable features but which do not satisfy all the constraints. The optimal paths up to each of the relaxed states will be obtained, using the normal recursive formula (equation 4.1). The relaxed states would only be allowed to persist for one or two time stages.

Role of the IKBS

Dynamic programming (DP) has several noteworthy features in contrast to other optimization methods – it is intuitively logical and appealing, it separates the optimization procedure from the calculation of the system states and because of this existing algorithms can be embedded in a DP program. However, the number of stages to be examined can grow very rapidly, often referred to as 'the curse of dimensionality'. Problems involving multiple interactions between variables are prone to this. A rule-based IKBS provides a means to restrict the number of possibilities worth examining, at the same time providing a framework for including the basic constraints such as system capacity, maximum capacity of a particular type of plant etc.

Typical rules, which can be classified as pertaining to the definition of individual states or transition between states, will be as follows:

1. total capacity between upper and lower limits related to the expected peak demand (transmission and load security effects could be incorporated by including similar rules for parts of the system), which may be quantified by the use of a loss of load probability (LOLP) or similar criterion;
2. maximum limits on the capacity of each type of plant which can be installed at each stage;
3. specific plant requirements, e.g. a unit of a certain type must be installed at a specific stage;
4. components of the generation mix must be within specified limits;
5. priority rules to specify the order in which units should be installed or sites developed;
6. reliability rules, e.g. LOLP less than specified figure (a relaxed criterion may also be included);
7. 'and' rules to cover cases where types of plant necessarily complement each other, e.g. each nuclear unit must have with it a gas turbine unit to provide secure house-service supplies;
8. 'or' rules to cover cases such as 'if site A is started then site B cannot be started';
9. 'nor' rules to cover cases such as a certain development not occurring in a specified time span.

Role of fuzzy logic

Some judgements based on expert opinion will be crucial in decision making but not possible to formulate in 'hard constraint' form, for instance a compromise between the cost of fuel and its supply security.

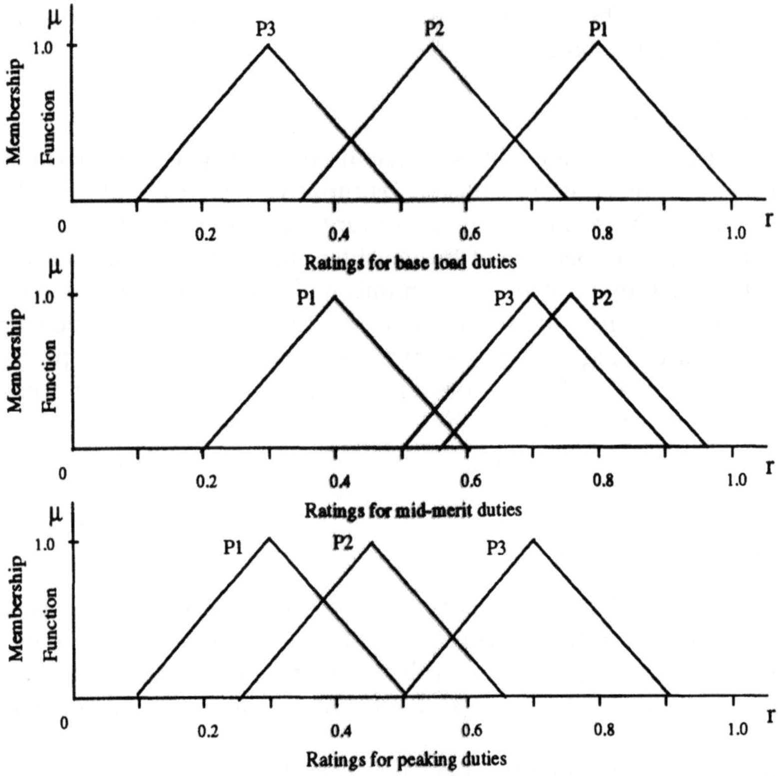

Figure 4.3 Templates of suitability of plant of different types for various duties.

The application of fuzzy logic [16] to this problem is described in [13]. If r_{ij} represents the rating (good, bad etc.) of the jth attribute of plant option i, its membership function will be m_{ij} (r_{ij}) and this can be selected from a pool of judgement **templates**. Knowing for any particular development strategy the capacity of plant option i, its contribution to each particular attribute can be calculated as a weighting factor

$$G_i / \Sigma_{i=1}^{n} G_i.$$

Taking as an example the need to provide a mix of plant capable of meeting the base load, peaking and emergency duties, the suitability of say three candidate plant types for providing these duties may be assessed by the planning and generation design staffs as in Figure 4.3; plant type P_1 would be typically base-load plant, P_2 mid-merit plant and P_3 peaking plant.

The second type of weights are attribute weights W_{aj} (j = 1 to m) representing the relative importance of the different attributes to final decision making. These are subjective judgements and, as with the attribute ratings themselves, are most appropriately represented by

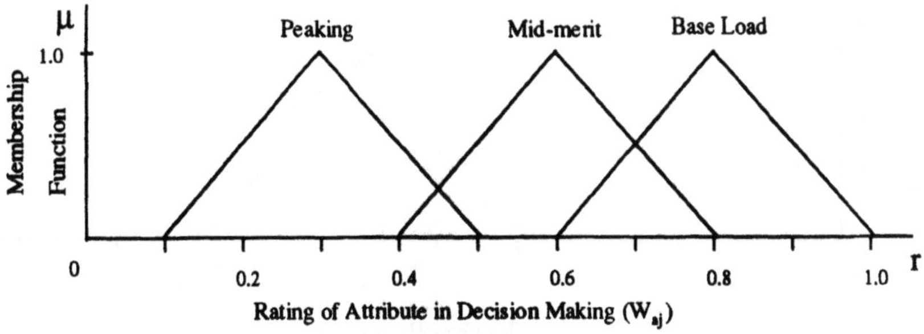

Figure 4.4 Importance of attributes in decision making.

fuzzy sets and variables, denoted by concepts such as 'very important', 'important', . . ., 'unimportant'. They are supplied as expert opinions by the program users. The program then makes an appropriate selection from the same set of judgement 'templates' referred to previously. (It will now be clear that these templates serve a general purpose function and a set having finer or coarser graduation may be selected or set up as appropriate to modelling requirements.) The second type of weights will constitute a fuzzy set $W_a = (W_{aj})$ where $W_{aj} = \{(W_{aj}, \mu W_{aj})\}$, and are independent of i as illustrated in Figure 4.4. The rating of a state with respect to attribute is fuzzy and is obtained by combining the contributions from the various plant options.

The final question is whether the state is acceptable or not with respect to the attributes. This can be done by comparing the state membership function, attribute by attribute, with the criteria or by forming a composite membership function with a criterion of minimum acceptable goodness as indicated in Figure 4.5.

(c) An example

A small example which demonstrates the proposals is included in David and Zhao [13]. The main features of the example are as follows.

1. Time horizon of study: 20 years (starting 1990).
2. Existing and committed generation at 1990: six generating units (oil, coal, hydro, gas turbine, ranging in size from 100 to 650 MW and operating cost 0.0013 to 0.026 \$/kWh; total generation capacity 2450 MW).
3. Peak system demand: 3000 MW in 1995 rising to 7000 MW in 2010, in defined five-year stages.
4. Future generation candidates: six candidate types of generation are specified: oil, coal, two sizes of hydro, nuclear, gas turbine. Sizes range from 100 to 900 MW, capacity costs from 250 to 1500 \$/kW, operating costs from 0.002 to 0.026 \$/kWh, forced outage rates from 1% to 5% and maximum numbers of units allowable between two and six.

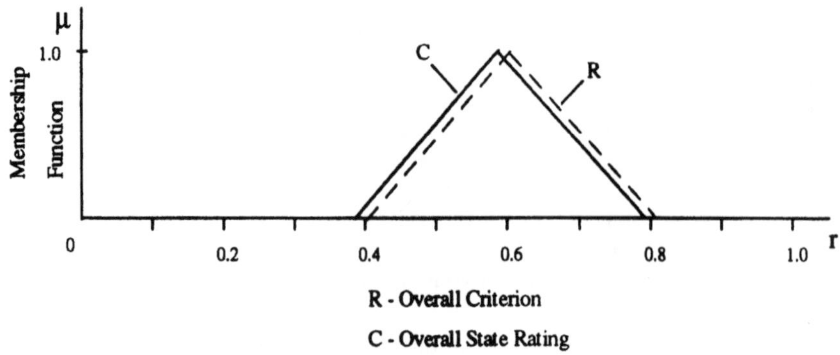

Figure 4.5 Suitability of plant mix.

5. IKBS rules – eight types of rules were used:
   (a) installation capability per stage (e.g. two oil units in stage 3);
   (b) total system capacity to lie between 1.1 and 1.4 times the peak demand;
   (c) order of installation (a strategy rule, one nuclear must be installed at stage 3);
   (d) oil and coal units must not be installed simultaneously at stages 1 and 2;
   (e) gas turbine capacity must be greater than 0.1 nuclear capacity;
   (f) smaller hydro must be installed before larger hydro;
   (g) fuzzy mix – membership values of loading capacity for base, peak and emergency duties must be greater than judged values specified by planners;
   (h) normal and relaxed loss of load probabilities are specified.

Numerous results are given demonstrating the progressive reduction in problem size (number of states) as rules are incorporated. Thus using the overall criterion and state rating approach, when the IKBS rules are excluded there are 51 states in stage 1 rising to 3876 states in stage 4. Including all rules except the fuzzy mix rule reduces these to eight and 105 states respectively. Adding the fuzzy mix rule further reduces these to six and 63 states, whilst testing the validity of the state on an attribute by attribute basis gives a final reduction to five states in stage 1 and 55 states in stage 4.

(d) Comment

The use of dynamic programming to explore plant extension programmes has been studied by various workers, including the author. However, the extensions put forward by Zhao David in the papers summarized here (an IKBS to formulate the logic of allowable generation expansions and restrict the area of search, backward chaining to prevent the blocking of useful strategies, and use of fuzzy logic) seem to show significant advantages.

4.8 DESCRIPTIONS OF SOME PROPOSED AND ACTUAL APPLICATIONS OF IKBSs IN TRANSMISSION NETWORK PLANNING

The two stages of network planning – determination of plant parameters, and design of networks using plant of these ratings – have been outlined in section 4.5.2. Proposals for the use of IKBS applications to these tasks in the transmission area will be described below.

Typically the design process, when considering the need and parameters for a new transmission voltage, will consist of the following steps.

1. Estimate the major energy flows through the system for various generation and demand scenarios.
2. Select a number of possible transmission voltages, line and transformer ratings etc. and cost point-to-point schemes for a range of power capacities and distances, guided by the requirements indicated in step 1.
3. Cost an outline network for the major energy flows from step 1, using the best alternatives from step 2.
4. Choosing the best alternatives from step 3, examine these in detail with respect to detailed plant design and costs; detailed system design (e.g. substation layouts); reliability; environmental impact, protection and control issues etc.
5. Choose the standards, e.g. rated transmission voltage; line configurations (single or double circuit); conductor configurations, e.g. twin, quad and cross-section; transformer voltages, winding configurations and ratings; switchgear normal and fault ratings; series and shunt reactive compensation; insulation levels; substation layouts for different functions (e.g. generating-station connection, major switching point, step-down connection to lower voltage networks), protection requirements, local and system control requirements etc.
6. Apply these standards in the initial establishment of the new transmission network and the year-by-year extensions.

The voltage standard would be expected to be used for general network development for 15 years or more although specific point-to-point needs could be dealt with by one-off designs; for instance, large remote generating stations could be connected by direct current or still higher voltage AC.

4.8.1 An IKBS for the preliminary design of power transmission networks

This work has been performed by staff at McGill University and Hydro Quebec, Montreal [8, 9] and has resulted in a package TRANSEPT combining IKBS modules and classical power system simulation tools. It can be used to explore alternating current or direct current developments.

(a) Objective and basic approach

The objectives of the project have been to explore the potential of IKBSs in the preliminary design of AC/DC power transmission networks, and to produce a planning tool for this purpose. Not least, the planners' expert knowledge can be retained and passed on to future staff. The approach broadly follows that outlined above, incorporating IKBS where appropriate.

The overall structure of TRANSEPT is shown in Figure 4.6 [9] combining data, knowledge base, simulation and costing. The object database includes data on equipment, design expertise, design criteria and housekeeping. There are seven rule modules, the principal ones being for design, simulation control, search strategy, costing and learning. Voltage control, load flow, transient stability and reliability are examined in the simulation models.

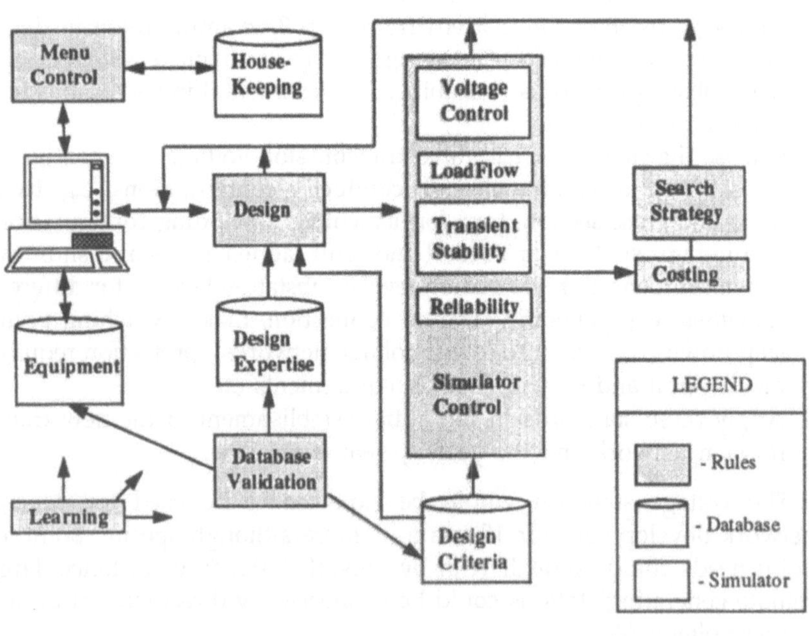

Figure 4.6 Overall structure of TRANSEPT.

(b) System design (AC system)

Design rules are applied to determine a skeleton structure for the point-to-point transmitted power and distance requirements. Heuristic rules provide recommended voltage ranges. It is anticipated that these values will be automatically refined by the IKBS using learning rules. The number of circuits is based on surge impedance loading for a given maximum angle over the compensated apparent line length. The number of intermediate substations is suggested. Reactive compensation is determined based on acceptable temporary overvoltages. Sensitivity analyses are performed around the base designs, providing information for later guided searches. A DC design module has also been developed.

(c) Costing

Costing rules enable capital, including spares, and operating costs, including unreliability and losses to be estimated.

(d) Search strategies

Voltage level, number of circuits and number of intermediate substations around the original design figures are examined. The strategies incorporate heuristic design rules.

(e) Learning

TRANSEPT estimates a least-cost design for the given requirements and hence could use its own results to extend and modify the design expertise database.

(f) Simulation control

The simulation modules provide system optimization as well as simulation functions, refining the skeleton structure determined by the design rules. Thus a transient stability simulation (equal area criterion) is used to estimate series capacitance and static VAR compensator parameters.

(g) Implementation

The inference engine is based on the IKBS shell Nexpert Object. TRANSEPT enables the planner to reason at a higher level and to examine the effects of different design criteria and constraints, studies which would otherwise be prohibitive due to computational or time constraints. The program is emerging as the centrepiece of a group of IKBSs covering the whole gamut of preliminary transmission planning.

(h) Example

Results obtained using the TRANSEPT package on the Hydro-Quebec network are given in Loud *et al.* [9], schemes to transmit 10 000 MW over a distance of 1000 km being examined. AC system voltages of 315 kV, 500 kV, 735 kV and 1200 kV and a DC 800 kV system were examined, several hundred alternatives in total. Results for one DC and the four cheapest AC schemes are given including costs with and without the reliability term and with and without spares. The expected reliabilities are shown.

4.8.2 Development of an IKBS for long-term planning of power transmission networks

In contrast to the previous project, this development, SINTIA II (interactive synthesis of transmission using artificial intelligence), deals with the expansion planning of a transmission network over a 20-year period [17]. An earlier system, SINTRA (transmission interactive synthesis) [18] included the logic for network extension. Its algorithms had been written in Fortran but it was difficult to change or expand the heuristics to adapt it to different network characteristics.

(a) Objective of SINTIA II

The goal of long-term transmission expansion planning for an established network is to determine what changes should be made to the transmission network and at what dates, in order to meet (at minimum cost under all credible conditions over the whole time span) the technical constraints of power flow and fault levels etc. The first development, SINTRA, provided an interactive tool to support this planning task. The main goals of SINTIA II were to provide flexibility and a modern user interface to SINTRA without losing the software investment in that system and the benefits it brought.

(b) Interactive planning

Using the agreed set of network voltages and plant ratings, the basic steps in determining a long-term expansion plan in SINTIA II are as follows.

1. Update demand and generation scenarios for each or selected years of the planning period.

2. Add lines to the network until any overloads are eliminated. A heuristic algorithm based on a linear (DC) load flow is used. The planner can also suggest additions manually.
3. Since any change in a network will have effects throughout the network, it is possible that a later addition may render unnecessary an earlier addition. In such cases early additions found to be unnecessary at a later stage can be eliminated by advancing the date of the later addition.
4. Examine the credible contingency conditions (e.g. single circuit, perhaps some double circuit outages).
5. Examine the resulting proposals in the context of those needed for the final year and repeat the process attempting to exclude any additions not needed in the final year.

(c) Linearized power flows and choice of new circuits for addition

As often in planning and operation studies, a linearized power flow is used. This gives a good approximation to active power flows in transmission networks (the reactance/resistance (X/R) ratios for circuits are high). The model is linear and, therefore, can be easily included in linear programming models [1]. Other advantages are that there are no convergence problems, computation times are decreased and configuration changes can be easily and rapidly handled.

It is a property of the DC load flow that the product P^2X (P being the circuit flow and X the circuit reactance) summed over all circuits is a minimum (the author once considered using this as the basis for solving a linear program). It is proposed in this model to use this property as a 'least effort criterion' to rank possible circuit additions, taking the change in the total system ΣP^2X product on the addition of a circuit, divided by the circuit cost, as a figure of merit for the possible new circuit.

(d) Application of IKBSs

As noted earlier, IKBSs were introduced to facilitate software development and maintenance, rather than to provide the basic logical or reasoning models. Three such applications are outlined.

Use of frames for data storage and manipulation

Frames, with slots for attributes within frames, are used to represent the various components of the network and also the network itself and elements dealt with in the planning process. Some examples are as follows.

1. Frame: possible new branch (e.g. line or transformer)
   Slots: branch number, terminal buses, voltage, type (line or transformer), cost parameters, least effort index etc.
2. Frame: existing branch
   Slots: branch number, terminal buses, technical parameters etc.
3. Frame: present network
   Slots: year, overloads etc.

As there was no frame representation directly available in Prolog, an **abstract data type** approach was used to obtain, partially, the desired frame functionality. In this the frame or slot is used to define and outline the required information from the Prolog database.

Programming

Programming languages such as Fortran and C are very suitable for numerical algorithms, but others such as Prolog and OPS5 are best suited to deal with symbolic knowledge about the network and overall control flow. In addition to Fortran and C, Prolog was used in both procedural and rule-based styles. It was found that there was a big reduction in source code when the planning control mechanisms were migrated from Fortran as used in SINTRA to Prolog in SINTIA II.

Coupling numeric and symbolic computing

One of the main characteristics of the SINTIA II program was the coupling of symbolic and numerical computing. A prototype version of the program made use of the compatibility between Prolog and C compilers for personal computers, some numerical Fortran subroutines from SINTRA having been translated into C for this purpose.

(e) Conclusions

In addition to the reduction in source code mentioned earlier, it was found that functions, heuristics and rules could be modified or added easily. The code was very understandable. The numerical subroutines from SINTRA could be reused unchanged with the software coupling used.

4.9 DESCRIPTIONS OF SOME PROPOSED AND ACTUAL APPLICATIONS OF IKBSs IN DISTRIBUTION NETWORK PLANNING

The stages in the planning of distribution networks are in principle the same as in the planning of transmission networks, namely plant standardization followed by detailed design for specific reinforcements. The distinction between the stages is however much clearer because of the need to standardize the parameters of plant for the very wide range of conditions found between local distribution networks in urban, industrial, suburban and rural areas. As regards the application of IKBSs, the author is not aware of any proposals for standardization, but there have been a number of proposals for their use in distribution network design [5, 19]. Two of these will be described here, one appropriate for high-voltage (HV) (say 6–25 kV) and the other for low-voltage (LV) networks.

4.9.1 EHV/HV substation location and HV network design [20]

(a) Objective

The objective of this work was to produce a tool for distribution planning engineers to determine optimal locations of EHV/HV substations and HV networks to supply a number of HV/LV local substations. The tool would contain engineers' heuristic rules and mathematical methods. The optimum locations would be those for which the total losses in the HV cable network would be a minimum.

(b) Method

Starting with a grid on which important physical features and HV substation locations are marked a simple IKBS and logic are used to determine possible EHV/HV substation sites and HV configuration:

A. Six rules are used to eliminate unavailable EHV/HV substation sites (e.g. the proposed site is located on a highway or is too expensive).
B. If the total load supplied on an EHV/HV substation exceeds 90% of its installed capacity then increase this capacity or reallocate some of the load points supplied to another EHV/HV substation.
C. The number of HV feeders from each EHV/HV substation must be between upper and lower limits which are dependent on the installed capacity of the substation (at least three and at most six feeders from a 30 MVA substation).
D. If the load on an HV feeder exceeds its assigned ruling, then some loads on this feeder must be reallocated to other feeders.
E. Not more than two load points may be connected to the same load point.

The IKBS was written in Prolog. The design process is as follows.

1. Calculate initial substation locations.
2. Iterate the feeder power loss/location calculation until the total HV feeder losses are substantially constant (see below).

3. Calculate the distances between load points and EHV/HV substations and allocate each load point to its nearest substation.
4. Repeat steps 2 and 3 until no further reallocation of load points is required.
5. Designating the grid square in which an EHV/HV substation is located as a substation square, connect the load points in that square to the substation (the number of feeders will thus equal the number of load points).
6. Check rule C above, including the nearest load points in adjacent grid squares or deleting the remote load points in the substation square as appropriate to satisfy rule C.
7. Connect each load point within neighbouring squares to the closest load point within the substation squares until all load points have been connected.
8. Rules D and E are checked in this process and appropriate action taken if violated.

In step 2 above, revised (X, Y) coordinates of the location of each EHV/HV substation are calculated as a function of the losses for the previous location, these losses being dependent on the inverse of the distances between the EHV/HV substation and its associated local substations.

(c) Example

The technique is demonstrated by its application to the design of a network to supply 23 load points, with demands ranging between 400 and 3500 kVA, to be supplied from two EHV/HV substations each of 30 MVA installed capacity. The output shows the locations of the EHV/HV substations (X, Y coordinates), the loads on these, the HV network power loss and the HV feeder configuration as indicated in Figure 4.7.

4.9.2 Connection of new customers [21]

(a) Objective

The objective of the authors was to develop a program to determine connections for new customers in a distribution network. As in other applications of IKBSs to distribution system planning, a relatively small and even simple task is considered, but one which occurs very frequently and overall involves large capital expenditures in total over the whole system.

EHV Substation Locations.

(x,y,co-ordinates)

Network Configuration.

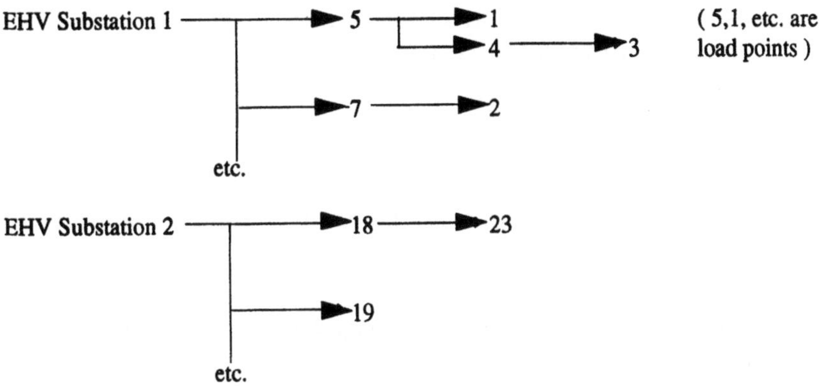

EHV Substation Loads

EHV Substation 1 = 25,600 kVA
EHV Substation 2 = 14,200 kVA

Total Network Power Loss = 647.98 kW

Figure 4.7 Location and connection of EHV/HV substations in a distribution network output.

(b) Method

A graphical map system (GMS), distribution system database and IKBS are integrated to produce a *'designer friendly'* approach. The IKBS incorporates the planner's rules and heuristic knowledge. The approach is as follows.

1. The planner calls the GMS and inputs information relating to the new customer (service region, customer location, contract capacity and forecast demand, service system (e.g. single phase) and conductor size).
2. The system responds with an appropriate area in which to search for possible HV/LV transformers to supply the new customer and examines the resulting candidate transformers to determine what capacity margins exist, and the alternatives of providing a new transformer or replacing an existing one by one of larger size. Transfer of demand between transformers is considered.
3. The locations of the candidate transformers are considered, the optimum solution being taken as the transformer nearest the customer.

4. The operations carried out by the inference engine are:
   (a) customer GMS data;
   (b) secondary line length condition;
   (c) change of transformer capacity;
   (d) decision on transformer location;
   (e) decision on load transfer between transformers;
   (f) decision on transformer capacity;
   (g) selection of circuit size.

(c) Implementation

The IKBS is written in Prolog, with the rule base comprising nearly 100 rules.

4.10 DESCRIPTIONS OF SOME PROPOSED AND ACTUAL IKBSs IN POWER SYSTEM ANALYSIS

Virtually all forms of steady state and transient (electromechanical) power system analysis are used in system planning studies. The relative number of applications is judged to be (from most to least):

1. active power load flow (the linear approximation, sometimes called DC load flow);
2. full (AC) load flow, including voltage levels;
3. balanced (three-phase) fault;
4. transient stability;
5. unbalanced (e.g. single and two-phase to earth) fault;
6. dynamic stability;
7. voltage stability.

At present most planning studies are essentially performed manually, with engineers proposing schemes and then testing if these schemes are adequate to provide secure and economic supply using the appropriate system analysis program. Sometimes optimum design and analysis programs are combined in one package.

System analysis studies require substantial amounts of data and often produce voluminous outputs (in broad terms, the further down the above list the more the input and output). IKBSs have been developed to assist in collating and checking data and in interpreting results. Sometimes, as in the examples described here, the IKBS is also used to optimize some aspect of the proposed system, in itself a mechanism which should enable the amount of output to be reduced.

The incentive to develop these IKBSs is to reduce the human resources needed for power system analysis, conserve knowledge in this area and hopefully improve the optimality of the solution.

4.10.1 Interactive load flow for power system planning

A paper describing 'an intelligent load-flow engine for power system planning' was published in 1986, probably the first in the field [22]. The objective was to provide a user-friendly and efficient planning environment incorporating advice to the user on changes to remove insecurities or bring variables within limits. The 'load-flow engine' included a graphic user interface, an editor to access and edit load-flow data, and a consultation module to interact with the IKBS.

The major rules for load-flow planning were, in summary, as follows.

1. If no circuit is overloaded and all voltages and reactive power generations are within limits, then the system condition is viable.
2. If a line is overloaded and generation (not nuclear) at the receiving end is not fully loaded, increase the generation by the lower of its margin and the amount of overload.
3. This rule, for generation at the sending end of the line, is the converse of rule 2.
4. If a bus voltage is above the limit, then switch in shunt reactors, or reduce reactive generation or lower the taps of transformers on plant near the bus with high voltage.
5. This rule, for bus voltage below limits, is the converse of rule 4.
6. If the reactive power output of generation is above the upper limit, then reduce the automatic voltage regulator (AVR) reference voltage.
7. This rule is the converse of rule 6.

The impedances between the bus at which the voltage is outside limits and the buses at which reactive power changes can be made are used to prioritize these adjustments. (C.C. Liu and others have demonstrated the use of sensitivity factors from the network Jacobian in conjunction with IKBSs in problems of voltage/VAR control (e.g. Ying-Yi and Liu [23]). The procedure is interactive – adjust, compute load flow, adjust etc. An example for a system of 201 buses and 285 lines required some 90 rules. The graphical interface was used to display the system diagram with problem areas highlighted and, in a separate area, the recommended actions to remove the problems. An explanation capability was provided.

Although there is a reference in the paper to including other forms of analysis to cover, e.g. transient stability and short circuit levels, it is not clear whether this was done. Nevertheless, the development was innovative and forward looking.

4.10.2 Support to load-flow calculations

Noting that load-flow computation is one of the tasks most frequently required in network planning, the Research Centre of ENEL (the Italian

Electricity Board) has developed an IKBS called SELF (expert system for load flow) to provide support to systems planners for load-flow calculations [24,25]. This system will be described in the following chapter (section 5.2.2) and so will not be expanded upon here except to provide a summary of some of the results obtained from its use.

The system has been tested on very large transmission networks (500–600 nodes and 800–1000 nodes). In all the tests appropriate solutions complying with the constraints were found. The time needed to analyse a 400-node network was of the order of a few tens of minutes (HP 9000/400 workstation) excluding the time taken for exploration.

Cicoria *et al.* comment that although solutions may not be optimal (e.g. one can note the exclusion of economic factors in some of the procedures proposed to rectify voltage and low flow conditions), and the computing times are sometimes longer, SELF provides detailed alphanumeric and graphical explanations of the proposals made [24,25].

4.10.3 Support for transient stability studies

Transmission requirements and problems have been increasing for decades as generation sites close to load centres have been developed, demands have increased and opposition to transmission lines has grown. This has meant that it has been necessary not only to increase the capability of transmission systems by such means as series capacitors, high-speed excitation, static VAR compensators, power system stabilizers, braking resistors, faster operating circuit breakers, but also to predict transient and other stability limits reliably.

The transient stability of a system will depend on the type of fault, fault location, fault duration and system loading condition (for instance, the transient stability of a system may be more critical at low load conditions). Each of these parameters must be selected to give a 'worst credible fault' condition, requiring either extensive testing of alternatives or considerable experience on the part of the system analyst. The problem is compounded by the fact that the standard step-by-step analysis models do not provide explicit margins of stability, only that the system is stable or unstable. Finally there is the problem of interpreting the swing curves – will the rotor swings die out or will the system go unstable. The need to minimize the call on experienced staff, provide support to these and even more to less experienced staff and to conserve expertise has led to work on the use of IKBSs to support stability studies.

(a) Transient stability IKBS

GE Company (USA) and Tokyo Electric Power Company (Japan) have made a major contribution to the development of IKBS concepts and prototype programs to support power system stabilities. The work is described in Akimoto *et al.* [26,27], the first paper covers transient stability in particular, and the second is an extension to incorporate small disturbance stability and training.

Facilities provided
The suite of programs in total addresses the problems outlined above, providing such features as (phase 1 [26]):

1. study organization and parameters;
2. time simulation;
3. output analysis;
4. identification of problem cause;
5. specification of further test cases;
6. concepts for remedial measures;

and (phase 2 [27]):

1. training;
2. steady-state stability;
3. remedial measures and power system design to improve stability;
4. scenario automation;
5. advanced program structure.

Structure
The structure of the transient stability IKBS is indicated in Figure 4.8 [26]. Some of the functions covered by the various modules are as follows.

1. **Study organization**. This IKBS assists in defining and checking the parameters for the study, for instance the initial system conditions and the potential contingencies. Component models and data are checked. The potential critical contingencies are selected on the basis of which generators have large rotor angles relative to the system centre of inertia and of circuits which have high surge impedance loading. These are run through a fast (direct method) stability screening program and stability margins evaluated for each system condition and contingency. The most critical cases (a key criterion is the ratio of critical clearance time to stated clearance time) are listed for evaluation by a step-by-step program.
2. **Time simulation (step-by-step analysis)**. A simplified transient stability program was used in the prototype development, but more extensive models were proposed for the future. An important feature added to the time simulation program was a check on run completion and a preliminary analysis of results ('validate execution').
3. **Output analysis**. This module assists analysis of the study results.

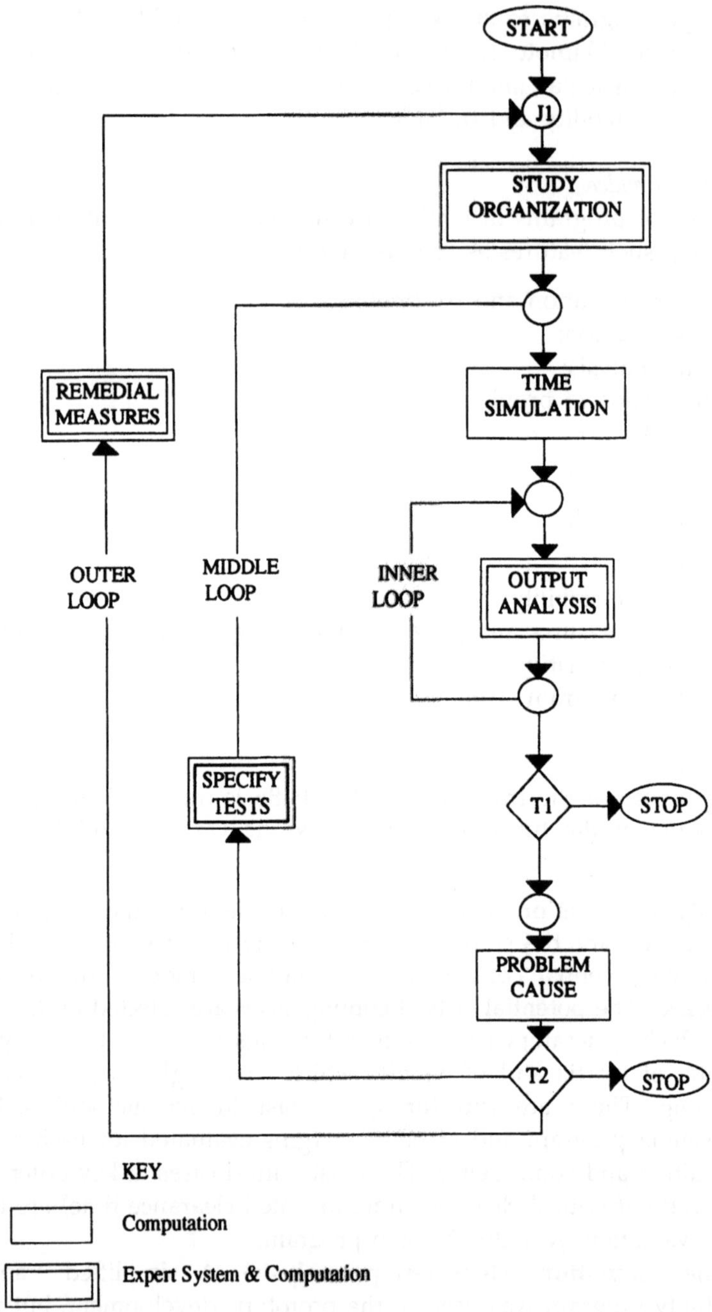

Figure 4.8 Framework for an expert system to support transient stability studies.

Features used to assess the stability of the system include the following.

(a) Are key variables approaching constant values within specified limits at the end of the run?
(b) Are the changes in key variables approaching zero at the end of the run?

4. **Problem cause analysis**. An IKBS was developed to assist in identification of the cause of instability, to specify further test cases and to suggest (conceptual only) remedial measures In this a **macro** instability classification defines the general type of instability, e.g. individual generator, coherent group or widespread instability. A **micro** classification relates instability to slow clearance, high post-fault impedance, low voltages or isolation of a part of the system.
5. **Specify test**. This helps the user to decide what further tests to run, using a direct method analysis of adjacent busbar and circuit faults to indicate the extent of system weakness. The sensitivities of the critical clearance times are also studied.

Facilities added in the second phase of the development [27] were as follows:

1. **Dynamic stability assessment**. The dynamic stability of the system is assessed using eigenvalue analysis plus rulebases, application software and graphics to support analysis of the eigenvalue results.
2. **Recommendations for remedial measures**. The IKBS models suggested measures to improve stability of the system from a list of seven possibilities, ranging through addition of reactive compensation, more transmission switching, power system stabilizers, reduced fault clearance times, and demand shedding.
3. **Design of remedial measures**. This appears to be a generalization of the remedial measures recommendation facility. For instance, alternative sites are included for the various remedial measures and 'adjust generation dispatch' is added.
4. **Training module**. A module was developed to train users in carrying out a transient stability study within an IKBS framework and to teach users basic concepts of stability analysis.
5. **Scenario automation (transient stability)**. This allows the user to achieve automatic execution of a selected sequence of processing steps.
6. **Advanced program structure (director program)**. The director program to provide overall execution control included in the first phase was revised in the second phase to include the new facilities and a relational database.

(b) knowledge based support system for power system analysis

Engineers of the Toshiba Corporation (Japan) have also reported work on the development of a knowledge based support system for power system analysis [28]. The incentive was to produce a system so that a nonspecialist, even a beginner, in power system analysis could handle the large volume of data, interpret results correctly and effectively analyse the power system performance at expert level.

Functions and structure

The main functions provided by the system are as follows.

1. Support is provided for production and testing of input data (checking for consistency and reasonable comparison with standard data). The checks include network configuration, nodal voltages and injections, generator, AVR and power system stabilizer constants, proposed fault conditions. The closed-loop stability of the AVR model is checked. System diagrams with load flows can be shown graphically on the human computer interface.
2. There is also support for checking output data and understanding phenomena. A check is made for load-flow convergence (the Newton–Raphson solution) and rules provided to find the cause of nonconvergence and countermeasures to this. Overloads and generator outputs are checked. The stability of the transient-stability solution is judged using the calculated data on angular and speed deviations of the generators. Abnormal voltage and frequency are indicated.
3. Modifications for incorrect data are suggested.

The structure of the system is shown in Figure 4.9. A 'parent/child' frame type database is used, the parent frames being node, branch, generators, AVR and fault condition. The child frames are appropriate combinations of the parent frames.

4.11 DEMAND FORECASTING

Reference was made in section 4.3 to the techniques used in long-term forecasting of electricity demand. This is clearly a diffuse subject, calling on a wide range of data, opinions and models and it may be because of this characteristic that very little, if any work, has been done on the application of IKBSs to the task (the reader may recall that IKBSs perform best in the detailed analysis of well-defined problems). Several reviews of actual or potential applications of IKBSs to power system planning and operations problems have appeared (e.g. references 29–40). Virtually nothing is reported on demand forecasting as normally understood in the planning context, say five years and more ahead.

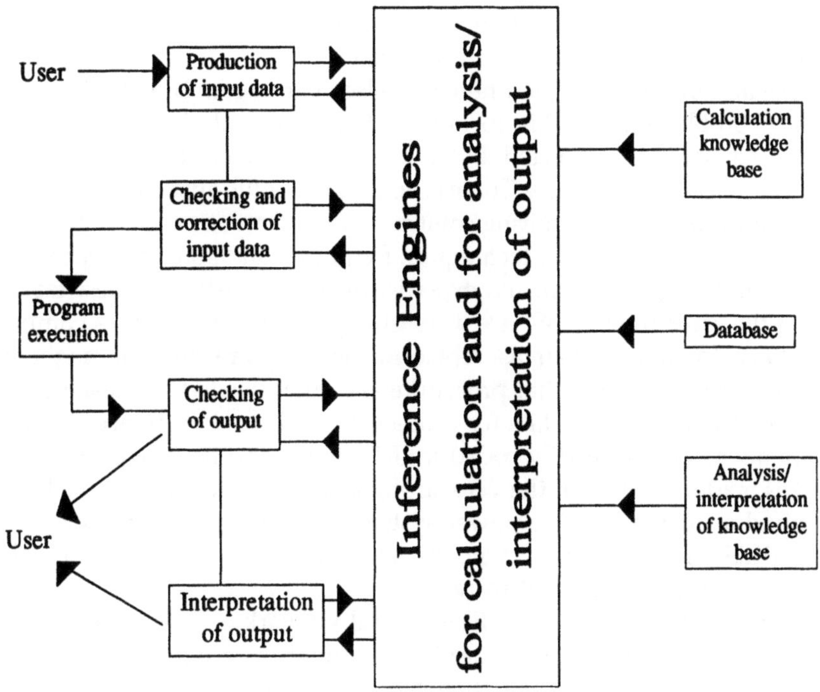

Figure 4.9 Structure of a knowledge based support system for power system analysis.

In contrast, numerous studies have been made on the application of IKBSs to short-term (minutes to days ahead) demand forecasting. This is a much more definitive problem and it is interesting to note that the frequent approach is to retain the algorithmic method, using an IKBS to provide a logical framework to define the parameters of the algorithmic model, including searching for similar past conditions.

4.12 TRENDS AND CONCLUSIONS

Artificial intelligence research has been pursued for nearly 50 years [40]. In the form of IKBSs, it began to penetrate into the mainstream of power system planning and operations in the mid to late 1980s. Thousands of papers have been published but even so the proposals for use fall into a few well-defined categories, the majority of which are in the system operation and control fields. It is indicated in [37] that based on a survey in 1990, long- and short-term planning applications in Japan accounted for some 11% of the total, and applications in control and monitoring some 86%. Judging from other reviews, these figures are probably representative of the situation worldwide. Some 12% of the proposed control applications were in practical use.

The author has suggested here and elsewhere a number of criteria by which problems amenable to and profitable for solution by the IKBS approach should be selected. By and large, surveys of proposed applications bear out this analysis. There is no doubt that the approach should be to look for problems requiring solutions rather than *vice versa*, in the context of existing solutions, techniques, frequency and spread of each task through the organization, and expertise of staff available for these tasks.

At the same time, IKBSs are increasingly being considered as complementary to, rather than as replacements for, conventional methods. Experience is showing that procedure and rules needed, for instance, to select particular sets of data for processing or alternative algorithms and parameters can be readily coded in IKBS terms and languages. Efficient use is made, thereby, of the very strong analytical and algorithmic background available to solve power system problems and of the power of the IKBS rule-based and symbolic approach, both in formulating the models and producing software.

Other AI techniques – artificial neural networks, genetic algorithms, fuzzy systems, simulated annealing (although it is debatable whether this is AI or another optimization procedure) are now being studied for potential applications in the power system planning and operations fields [41–44]. Decision trees and pattern recognition are also being revisited. The author suggests that each of these will be found to have its niche, defined not only by the problem type but also by organizational issues. Just as algorithmic and IKBS approaches complement each other, so may AI techniques in general complement each other [45], often in conjunction with algorithmic approaches.

REFERENCES

1. Knight, U.G. (1972) *Power Systems Engineering and Mathematics*, Pergamon Press.
2. Bennet, J.M., Dakin, F.V. and Knight, U.G. (1952) Digital computers and their application to some electrical engineering problems, Paper 304, *CIGRE*.
3. Massé, P. (1959) *Le choix des investisments*, Dunod, Paris.
4. Davies, C.H. (1982) *Scenarios and Electricity Demand*, CEGB Proof of Evidence to Sizewell B Power Station Public Enquiry, November.
5. Knight, U.G. (1991) Expert systems: part 1 – planning and operation of power systems; part 2 – applications in generation transmission and distribution, in *Decision Systems International*, Monaco.
6. *WASP-III Guidebook*, International Atomic Energy Agency, 1984.
7. *Electric Generation Expansion Analysis System* (5 volumes), EPRI Report EL-2561, 1982.
8. Galiana, F.D. and McGillis, D. (1988) Design of a longitudinal ehv transmission network: a knowledge based approach, in *First Symposium on Expert System Applications to Power Systems (ESAP)*, Stockholm, August.

9. Loud, L., Galiana, F.D. and McGillis, D. (1993) An integrated knowledge based system for the preliminary design of power transmission network, in *11th Power Systems Computation Conference (PSCC)*, Vol. 2.
10. Farghal, S.A. *et al.* (1988) Generation expansion planning: an expert system approach. *IEE Proceedings*, **135**, Part C, July.
11. David, A.K. and Zhao Rong-da (1989) Integrating expert systems with dynamic programming in generation expansion planning. *IEEE Transactions on Power Systems*, **4**(3).
12. David, A.K. and Zhao Rong-da. Advances in AI applications for power system planning.
13. David, A.K. and Zhao Rong-da (1991) An expert system with fuzzy sets for optimal planning. *IEEE Transactions on Power Systems*, **6** (1).
14. Bellman, R. (1957) *Dynamic Programming*, Princeton University Press.
15. Bellman, R. and Dreyfus, S.E. (1962) *Applied Dynamic Programming*, Princeton University Press.
16. Zadeh, L.A. (1965) Fuzzy sets. *Information and Control*, **8**.
17. Nasser, F.E.N. *et al.* (1989) Development of an expert system for long term planning of power transmission networks, in *2nd Symposium on Expert System Applications to Power Systems (ESAP)*, Seattle, USA.
18. Monticelli, A. *et al.* (1982) Interactive transmission network planning using a least-effort criterion. *IEEE Transactions on Power Applications and Systems*, **101**(10).
19. Knight, U.G. (1990) Knowledge based expert system and electricity distribution, Central Board of Irrigation and Power, New Delhi and Asian Development Bank, Regional Seminar on Computer Aided Distribution Management, Hyderabad.
20. Yuan-Yih Hsu and Jiann-Liang Chen (1990) Distribution planning using a knowledge based expert system. *IEEE Transactions on Power Delivery*, **5**(3).
21. Kim, H. *et al.* (1989) An expert system for the electric power distribution design system, in *IFAC Symposium on Power Systems and Power Plant Control*.
22. Fujiwara, R. *et al.* (1986) An intelligent load flow engine for power system planning. *IEEE Transactions on Power Systems*, **1**(3).
23. Ying-Yi, H. and Liu, C.C. (1992) A heuristic and algorithmic approach to VAR planning. *IEEE Transactions on Power Systems*, **7**(2).
24. Cicoria, R. *et al.* (1990) SELF, an expert system application to power system planning, in *Proceedings 10th International Workshop on Expert Systems and Their Application*, Avignon, France.
25. Cicoria, R. *et al.* (1994) SELF, an expert system for supporting the verification of network operating constraints in power transmission planning, in *Intelligent System Applications to Power Systems 94*, Montpellier, France.
26. Akimoto, Y., Price, W.W. *et al.* (1989) Transient stability expert system. *IEEE Transactions on Power Systems*, **4**(1).
27. Akimoto, Y., Price, W.W. *et al.* (1991) Application of expert system technology to power system stability studies, in *Third Symposium on Expert Systems Applications to Power Systems*, Tokyo.
28. Taguchi, Y. and Uemura, Y. (1991) knowledge based supporting system for power system analysis, in *Third Symposium on Expert Systems Applications to Power Systems*, Tokyo.
29. Knight, U.G. (1987) Expert systems in power system planning and operation from the viewpoint of a utility engineer, in *Power Systems Computation Conference*, Lisbon.

30. Tamura, Y. *et al.* (1988) An international survey of the present status and the perspective of expert systems on power system analysis and techniques, *CIGRE Study Committee 38*.
31. Zhang, Z.Z. *et al.* (1989) Expert systems in electrical power systems – a bibliographical survey. *IEEE Transactions on Power Systems*, **4**(4).
32. Liu, C.C. (1990) Knowledge based systems in power systems application and development methods. *Transactions of IEE Japan*, **110B**(4).
33. Tamura, Y. and Sasaki, H. (1991) Expert systems applied to voltage and var control. *Electra*, No. 139.
34. Galiana, F.D. *et al.* (1992) Expert systems in transmission planning. *Proceedings of the IEEE*, **80**(5).
35. Balu, N. *et al.* (1992) Review of expert systems in bulk power system planning and operation. *Proceedings of the IEEE*, **30**(5).
36. Germond, A.J. and Niebur, D. (1992) Survey of knowledge based systems in power systems: Europe. *Proceedings of the IEEE*, **30**(5).
37. Tanaka, H. *et al.* (1992) Research and development in expert systems applications to power systems in Japan. *Electrical Power and Energy Systems*, **14**(2/3).
38. Liu, C.C. and Van Son, P. (1993) Expert systems: development experience and user requirements. *Electra*, No. 146.
39. CIGRE Task Force 38.06.03 (1993) Practical use of expert systems in planning and operation of power systems. *Electra*, No. 146.
40. Oakley, B.W. (1987) Expert systems: artificial intelligence research comes of age, Public lecture at the Royal Society, London, May.
41. Nara, K. *et al.* (1996) A Tabu search algorithm for determining distribution tie lines, in *Proceedings of the International Conference on Intelligent System Applications to Power Systems (ISAP'96)*, pp. 266–70.
42. Michalik, G. and Mielczarski, W. (1996) Modelling of energy use patterns in the residential sector using linguistic variables, in *Proceedings of the International Conference on Intelligent System Applications to Power Systems (ISAP'96)*, pp. 278–82.
43. Parlos, A.G. *et al.* (1996) Development of an intelligent long-term electric load forecasting system, in *Proceedings of the International Conference on Intelligent System Applications to Power Systems (ISAP'96)*, pp. 288–92.
44. Muto, S. (1996) A special period load forecasting method based on order relations, in *Proceedings of the International Conference on Intelligent System Applications to Power Systems (ISAP'96)*, pp. 120–25.
45. Kagan, N. and Oliveira, C.C.B. (1996) A fuzzy constrained decision planning tool to model uncertainties in multiobjective configuration problems, in *Proceedings of the International Conference on Intelligent System Applications to Power Systems (ISAP'96)*, pp. 271–75.

5

Design of electrical plant and systems

Stefano Massucco

5.1 INTRODUCTION

5.1.1 General overview

Although the first significant applications of artificial intelligence in power systems have dealt mostly with diagnostics, monitoring and control, a certain interest in planning and design can be deduced from international reports [1,2].

Power system complexity, for both large and industrial systems, has increased greatly in recent years: the rapid growth of electrical consumption and stringent economic constraints are dictating that systems must be better planned, designed and operated. This ensures that every system component is best used, overall system performance is satisfactory and there are no adverse effects on security or excessive stresses on materials.

A greater demand for flexibility is one of the main characteristics of the modern power system [3]. Flexibility can be obviously obtained by adequate control and monitoring systems and, to this end, the rapid development of electronic and computer applications is providing improved facilities. Nevertheless, it is widely held that an important role is still played by design and planning.

For instance, in the design of an electrical system for industrial distribution, a good one-line diagram organization and configuration, and good voltage-level selection and load allocation are still of great importance to successful system operation. No significant advantage can be gained from the use of sophisticated control devices if the system has not been correctly designed.

When performing power system studies, data is not always well known or might only be known approximately. It is not always possible to define an adequate mathematical model of a physical problem, and the complexity of the problem limits the computational capability of tools used for analysis. Nevertheless there may exist people, with the appellation of 'experts', who are able to solve the problem. Sometimes these experts have difficulty in explaining their solution process because it is mainly based on either experience acquired over many years, very simple but efficient procedures or on heuristic assumptions and 'rules of thumb' [4]. In summary, some problems cannot be translated into the form required by a computer study, at least in the traditional sense, because their solution is typically a product of a human activity, namely problem solving.

Planning and design are aspects of engineering studies which mostly require the aforementioned abilities of a human being. In this sense, the application of 'artificial intelligence', and more specifically of 'intelligent knowledge based systems' (IKBSs), seems to be particularly appropriate.

Before examining existing applications of IKBSs in power system planning and design in more detail, a point, which is also made in the analysis reported in Dillon and Laughton [4] under the title 'international survey of expert system application', should be highlighted. That is, that these applications received less attention from the international community than monitoring and control, because of the large quantity of work necessary to assemble the knowledge required to make correct planning and design choices. On the other hand, the rapid evolution and development of tools for aiding knowledge acquisition from human experts, has encouraged researchers from university and industry to develop IKBS applications in the field of power system planning and design.

It is to be stressed that most of the present-day applications are at the prototype stage or can only be regarded as feasibility studies. Presently they cannot be regarded as complete alternatives to traditional solutions, but they can be used as beneficial aids for improving existing solutions.

5.1.2 Reasons for the application of IKBSs to design and planning

Before going into more detail, the two applications we are considering, namely planning and design, need to be better defined. The two subjects deserve to be considered jointly because of the important interaction there exists between them.

Planning is one of the basic human activities, and is related to defining an efficient sequence of actions in order to reach a given aim or goal. This definition would include determining the solution process of a problem without analysing the methods of solution in great detail. Planning generally involves subdividing a problem into smaller ones, suggesting possible conflicts among sub-problems, and also accounting for economic criteria.

Deterministic methods such as optimization have been successfully applied to planning problems. Constraint formulation must, however, be very detailed, and may lead to the overloading of the solution process, thus preventing a relatively quick decision. Moreover, goal redefinition is quite difficult due to the lack of flexibility in these approaches. IKBSs appear to be of interest for characterizing planners' experience in the form of heuristic rules rather than requiring a full mathematical or numeric formulation.

Design is the action of carrying out something in accordance with a given plan, and is above all the result of resolving conflicting requirements. Power system design in particular is strongly interrelated with other engineering disciplines. It requires different experts to co-operate in the solution of a problem. In power plant design for instance, mechanical engineers, thermal engineers, piping engineers, electrical and electronic engineers, and layout and civil engineers are required to meet the technical and economic requirements contained in the power plant contractual documents. Quite often, these experts' knowledge is scarcely formalized. Moreover, a lack of information can be experienced, at least in the initial design stage. In any case, it is difficult to coordinate and formalize the large quantity of information related to such a project.

According to these considerations, the area of power system design is one in which expert systems can be applied with significant benefits arising. This is due to the fact that design practices are the result of a compromise between several criteria including cost, reliability, stability and the environment, which cannot always be mathematically described [4].

With reference to a general paper about knowledge systems for planning support [5] and with specific reference to power systems, the most significant benefits to be expected from the application of IKBSs to design and planning are as follows.

1. It combines a flexible collection of system knowledge from different specialists.
2. It allows a systematic approach to the acquisition of design knowledge, in order that no significant information is missed or misused.
3. It allows the examination of different planning or design strategies, the modification of assumptions to ensure that no technical or economical constraints are violated, and for this to be done more quickly than permitted by traditional solutions and with a lower additional monetary cost.
4. It makes the understanding of design choices easier.
5. It improves the transmission of technical knowledge from the few highly qualified individuals to inexperienced engineers.

6. It forms an advanced instruction tool for training technical staff in the areas of both design and operation.
7. It aids the writing of proper documentation and manuals at different stages of project development, keeping technical information coherent and making its consultation more efficient. This applies not only for design purposes, but also for operation and maintenance.

While the first few items in the above list were analysed in the earlier considerations, knowledge transmission and instruction deserve further comment. This is an area where IKBSs could provide a significant improvement in system design [6].

One reason for knowledge conservation is the high turnover rate engineering companies have in expert designers. There is a great need to retain design experience accumulated over many years of work and make it transferable to young engineers.

Another consideration in favour of applying IKBSs is the possibility of providing senior engineers with a tool for updating their knowledge. It is commonly felt that every kind of project necessitates the stringent updating of technical and scientific knowledge at a very high rate. Expert systems could be the means of providing intelligent computer-aided instruction. The suggested international acronym for such an application is ITS (intelligent tutoring systems). More information on the subject can be found in Wenger [7].

5.2 THE STATE OF THE ART IN IKBS APPLICATIONS FOR THE DESIGN OF ELECTRICAL PLANT AND SYSTEMS

5.2.1 Introduction

The aim of this section is to provide an overview of IKBS applications for the design of electrical plant and systems. The word design is used here in its wider sense, in order that some typical problems in the inherently connected area of planning can also be addressed. Some interesting papers concerning the subjects will be examined. The extent of interest in IKBS applications for power systems is revealed through a number of of international symposia on the subject [8–12]. Some of the references, relating to design applications presented in this section, are taken from the proceedings of these symposia.

It should be noted that the papers cited do not constitute an exhaustive list. They were chosen because they were considered to be significant applications. Rather than performing a deep bibliographical analysis about a very specific design subject, the purpose of this section is to give an idea of potential applications of IKBSs in different design fields. The following subdivisions were adopted in collecting and presenting the selected papers:

1. applications concerning large system design and planning [13–16];
2. applications concerning the design of systems of limited dimensions, such as power plant, industrial systems, substations, and distribution systems [17–21];
3. applications for specific design problems regarding subsystems or equipment, such as electric cable design, grounding design and protection system design [22–25].

A short description of each reference follows.

5.2.2 Applications to large systems

The first interesting paper in this group deals with the development of an IKBS which provides preliminary designs for a point-to-point EHV transmission system [13]. The IKBS has been built with AC/DC transmission systems, for carrying power from a remote source to load centres, in mind. The following aspects of the design process are examined:

1. design parameters (e.g. power generation, line length, equipment cost and data, voltage and VAR limits, stability criteria, environmental factors etc.);
2. design variables (e.g. the number of parallel circuits, the voltage level, the location of series and shunt compensation, the generator data, details of converters and filters);
3. performance criteria (cost, security, reliability, robustness).

The IKBS searches for a solution in terms of voltage level for each transmission line, conductor size, and location of compensation. Each proposed design can be reviewed by varying some parameters. A user of the IKBS can adopt a previously stored design or can compare his or her own choice with some previously found solution in order to derive an improved design. The ability of the IKBS in 'identifying and representing in a systematic manner the knowledge needed to arrive at an acceptable design' was judged [13] to be one of the most significant benefits. The IKBS was developed by using the Gold Works expert system shell, and runs on a 386-based computer.

A second interesting application in large system design and planning is an IKBS for power system planning developed by ENEL (the Italian Electricity Generating Board) in co-operation with CISE [14]. The first aim of this IKBS, named SELF, is to assist a planning engineer in obtaining convergence of the iterative load-flow algorithm, and to help him or her determine solutions matching voltage and reactive power constraints.

This application is also a significant example of integrating IKBS and conventional software for power system analysis. Integration is one of the most important problems which must be correctly solved in the near future in order to fully exploit the potential of IKBSs without losing the benefit of the significant amount of work performed by power system researchers in recent decades [15].

The IKBS SELF is targeted at high-voltage network planning. It is interfaced with a large database containing alphanumeric and graphic network data, and has been applied to systems of more than 400 nodes. Other interfaces are provided for interaction with computer programs performing the load-flow, short-circuit and sensitivity studies.

SELF acts like an expert human planner in that it gets input data of the network under study, activates the load flow, and, in accordance with the computation results, decides which of the following must be performed: a convergence analysis, a violation elimination or a compensation. The authors of reference 14 consider SELF, which was developed in Lucid Common LISP, to be a valuable tool for giving significant time savings, for improving planning solutions and for training inexperienced planners by directing them through the decision process.

A third application [16] concerns an IKBS for insulation coordination, and is somewhat related to that of reference 13. In fact, the authors regard their IKBS as the second of a series of tools useful for producing an integrated package for power system planning. The present IKBS arises from the results in reference 13, and determines the level of insulation for power system equipment. The process of choosing a correct level of insulation is more of an engineering problem than a mathematical one, with decisions based on past experience and testing. The design solution must balance electrical stresses caused by overvoltages affecting the power system, and must also determine the strength necessary to front them. An adequate margin must be chosen to guarantee proper insulation. The IKBS is required to generate design parameters for insulation such as insulator length, clearance to ground, surge arrestor ratings and withstand voltages. The IKBS approach seems to be particularly well suited to the problem. The same hardware and software environment as in reference 13 was adopted.

5.2.3 Applications to systems of limited dimensions

Among this second group of applications, some efforts have been directed towards the design of power stations, substations and industrial systems. The IKBS ASDEP [17] was developed for the purpose of designing the electrical auxiliary system of a power plant. ASDEP (Auxiliary System Design and Evaluation Program), was written in LISP and Pascal. Its goal is to design a good electrical auxiliary system from a proposed initial design. The alternative approach of simplifying a highly sophisticated initial design was rejected as it did not reproduce the actual design process adopted by human experts.

The proposed initial design is a one-line diagram of the electrical system. The IKBS is able to improve the design by applying a group of design-change rules. A final section is included in ASDEP for evaluating and refining the current design. The refining section is based upon physical and reliability constraints.

A similar application was developed for addressing the problem of designing the electrical system of a power plant [18]. In this case, however, the principal aim of the IKBS was to develop the topology of the electrical network and to correctly allocate electrical loads to distribution busbars. The initial diagram submitted to the IKBS is improved by using rules contained in the knowledge base. The application [18] made use of a version of the LISP language, a commonly adopted AI language. The approach of using an IKBS shell rather than an AI language holds promise due to the facilities offered by these tools. This approach is presented in section 5.3, and the design of the electrical system by using an IKBS is analysed in greater detail. Information relating to the problem under examination, knowledge acquisition and the design methodology will be reported and commented on.

Another paper [19] presents an application of IKBSs to integrate design choices for an industrial system, namely the quantity and location of transformers, reactive compensation and the unification of network elements. Since the auxiliary system of a power plant can be somewhat viewed as an electrical industrial system, this last application can be thought of as the third example of this subject. This underlines the interest in IKBS applications for the auxiliary system design problem.

Another interesting paper reports on an AI application to distribution substations [20]. A software tool, which can be considered an IKBS, was developed using the Prolog language. This tool firstly generates all possible solutions for allocating loads to the busbars of distribution substations. It then successively identifies invalid solutions, and produces feasible solutions that match busbar constraints. These constraints may include connection constraints according to load type, ensuring that transformer or cable ratings are not exceeded etc. The hardware used for implementation was a personal computer, and applications to two 11 kV substations are reported.

The final example relates to a design aid for capacitor placement in distribution feeders [21]. Shunt capacitors are placed along feeders for regulating voltage, for power factor improvement and for the minimization of energy losses. The purpose of the developed prototype was to determine the location and size of capacitors. The IKBS rules have been built using OPS5, a computer language for knowledge base programming, and operate on a minicomputer.

5.2.4 Applications to the design of specific subsystems or equipment

The first example of IKBSs applied to the design of subsystems or equipment concerns the design of an electrical cable system [22]. The design process of an electrical cable system is a complex problem, whose solution must consider electrical, mechanical and thermal aspects as well as wiring regulations contained in national and international standards. The result of the design process is in terms of cable sizing and requires a multidisciplinary approach, mostly based on experience. This approach can be used for the cable sizing of electrical installations in industrial and commercial buildings, although this specific example is applied to commercial buildings.

A LISP-based IKBS shell called GOLDWORK II has been used on a personal computer to realize this IKBS which selects the type and size of cable to be used in a particular application. Starting from the general characteristics of the building and the number and purposes of the electrical circuits, the design process performed by the IKBS determines the current to be carried by the cables, chooses protective devices, and selects the type of cables and the cross-sectional area of cable conductors taking into account the method of installation and the grouping factors. It also checks for violation of the voltage-drop limit under normal conditions.

A second application [23] concerns an IKBS for AC/DC substation grounding. The IKBS has been developed using a shell called RULEMASTER and with reference to the relevant IEEE Guides and Standards. The IKBS is divided into two parts: the first concerns the theoretical aspects and the definition of a grounding system design; the second part is a design tool which provides useful help for installing and evaluating grounding systems. The IKBS makes use of external routines for numerical computation which have been written in Fortran or C. Two applications of this IKBS have been presented. The first example determines the size of grounding conductors, and the second calculates the current contribution of each line of a grounding grid.

The third application to be considered here is a design support system for protection coordination in industrial systems [24]. The IKBS was developed with the aim of automatically coordinating and selecting overcurrent protection devices within an industrial distribution system. It is widely known that the setting of a protection system is one of the most important and difficult problems to confront an electrical system designer, and that it is one of those fields in which experience plays a significant role. Consequently the application of IKBS technology seems to be highly appropriate. This IKBS was developed in LISP. It acquires circuit configuration data, performs short-circuit calculations, gets proper selection rules, searches for adequate protective devices from a database, verifies the coordination between upline and downline protection devices, and produces a final report on selected devices.

Another application in the protection-setting domain is presented in [25]. This example deals with transmission line protection, and has been developed in OPS83, a supporting tool for IKBSs. The proposed IKBS performs the selection of the protective devices for a transmission line, and the determination of optimal current transformer turns ratio and relay setting. An illustrative example, which refers to a portion of an existing electrical system, is presented.

By now the reader should have an idea of the potential application of IKBSs to power system design. The next section analyses a specific design problem in detail. Consideration of the expected benefits and future trends in the application of IKBSs to design follows in the final section of this chapter.

5.3 AN IKBS APPLICATION: ELECTRICAL AUXILIARY SYSTEM DESIGN IN A POWER STATION

5.3.1 Problem formulation

In order to test the capability of AI to solve a design problem, an existing design of a large thermal power station has been considered. Specifically, the project under consideration relates to a standard nuclear power station [26] of the Italian Electricity Generating Board (ENEL). This design was chosen because no great difference exists between its electrical system and that of a conventional fossil thermal unit, with the evident exception of the more stringent requirements placed on the emergency systems of the nuclear plant. The examined system would appear to be a significant test for developing an IKBS applied to a design problem, and is also capable of satisfying the basic requirements for a prototype approach, which can be expressed as follows [27].

1. The problem being faced is of limited dimensions.
2. Human experts in the field are available (few experts are better than many experts for the sake of avoiding conflict during knowledge acquisition).
3. There is the possibility of comparison with the procedures usually adopted.

The following two subsections will explain the application domain of the IKBS. The first will give a description of the design process adopted by a human expert when dealing with the electrical auxiliary system of a power plant. The second subsection is a description of the electrical system of the plant under consideration.

Although it is common practice to present a proposed method before a test example, it seemed worthwhile to adopt the opposite convention here, considering that many readers are more familiar with power system problems than with IKBS techniques. It is hoped that the reader will find it useful to be introduced to the expert system approach after the specific application has been described. Moreover this order is very similar to the interaction between the knowledge engineer (the individual responsible for filling the knowledge base) and the expert (the human designer). The transfer of information is often performed by means of interviews in which a description of the problem is firstly stated and subproblems identified, and the knowledge about the problem improved by successive meetings.

The following description can be viewed as the general description of the electrical auxiliary system design process which would allow the setting-up of the knowledge base.

(a) Design methodology

The problem to which the IKBS has been applied is the design of the electrical auxiliary system of a large thermal power station. The amount of power required by the whole auxiliary system is within 10% of the total power produced by the plant. The availability and reliability of the auxiliary system is of primary importance to both the safety of the plant and its ability to satisfy the power demand.

The design stages to be considered in the present application are the preliminary and conceptual designs, in which 'the station service auxiliary system is examined in terms of required electrical performances, economics and future plant needs. Auxiliary system design criteria are established, voltage levels determined, major equipment ratings selected, . . . ' [28].

This application will not provide a detailed design, but it is intended to give guidelines for the correct definition of the one-line diagram and for the load allocation to busbars. These guidelines are given in accordance with the procedure adopted by an expert human designer, and include: choosing the correct number of busbars to satisfy availability and reliability requirements, adopting separation criteria for design safety; proper choice of busbar voltages; selecting the correct types of transformers etc.

The starting point for the design process is a load list, that is, a collection of information, often organized in a database, concerning every piece of equipment in the power plant requiring an electrical supply. An example of an actual load list is given in Figure 5.1, which relates to the aforementioned Italian plant [26] at the preliminary stage of the project.

ELENCO UTENZE ELETTRICHE													
IDENTIF. UTENZA EL. 1 UTENZA RIDONDANTE 2 UTENZA RIDONDANTE 3 UTENZA RIDONDANTE	LOCALE	DESCRIZIONE UTENZA ELLETTRICA	CA MA CS	SD TS FF	MPT BHT VAL	PF1 PF4 PF7	PF2 PF5 PF8	PF3 PF6 PF9	SA1 SA2 SA3	SM TU	TIF SDI	QUADRO CAVO PENETR
15415 CA 0500A 0000	L1E000	POMPA PRINCIPALE	1E		1000,0	900,0	900,0	900,0	S30	0		
15415 CA 0700C 0000		TRENO A CCW	P	C	900,0	900,0	900,0	750,0	S30	MP		
		NOTA 1	A	TA	6000	750,0	900,0	900,0	S30			
15415 CA 0700C 0000	L1E000	POMPA BACKUP	EE		1000,0	900,0	900,0	900,0	S30	S		
15415 CA 0500A 0000		TRENO A CCW	P	C	900,0	900,0	900,0	750,0	S30	MP		
		NOTA I	C	TA	6000	750,0	900,0	900,0	S30			
15415 CA 6500B 0000	L2E000	POMPA BACKUP	1E		1000,0	900,0	900,0	900,0	S30	O		
15415 CA 6700D 0000		TRENA B CCW	P	C	900,0	900,0	900,0	750,0	S30	MP		
		NOTA 1	B	TA	60000	750,0	900,0	900,0	S30			
15415 CA 6700D 0000	L2E000	POMPA BACKUP	EE		1000,0	900,0	900,0	900,0	S30	S		
15415 CA 6500B 0000		TRENA B CCW	P	C	900,0	900,0	900,0	750,0	S30	MP		
		NOTA 1	D	TA	6000	750,0	900,0	900,0	S30			
15415 FF 0702N 0000	L1E000	VALV ISOLAMENTO	OE		3,20	3,20	3,20	3,20	00			
		POMPA CA0700C		O		3,20	3,20		00	VM		
			N	TA	380				00			
15415 FF 0801N 0000	L1E000	VALV ISOLAMENTO	OE		3,20	3,20	3,20	3,20	00			
		POMPA CA0700C		O		3,20	3,20		00	VM		
			N	TA	380				00			

Figure 5.1 Partial example of a load list at the preliminary design stage.

Typical information for each load in the list follows: power requirements at different operating conditions of the plant (such as plant start-up, plant shut-down, full power operation etc.); an alpha-numeric code to uniquely identify the equipment; a safety class; a description of the performed function; emergency power supply requirements. The efficiency of auxiliary motors, power factor and other electrical data complete the information set. The load list is maintained during the whole project history, and its updating is vital for the correct development of the design and also for the construction and operation of the power plant.

The result of the first step in the design process is the preparation of a possible configuration of the electrical scheme supplying power to the station auxiliaries, and this is used for more detailed studies. The next steps performed by a human expert are short-circuit, voltage-drop and motor-starting calculations. Due to a lack of information in the initial stages of the design process, these calculations are often performed by means of simplified procedures which can also use preliminary data instead of detailed data. Once a one-line diagram has been defined, technical specifications for the acquisition of main equipment, such as generators, transformers and large motors, can be written. At the final design stage, when actual equipment data sheets are available, design refinements are possible. Verifying calculations and the setting of protection devices can then be carried out. At this stage, the previously adopted simplified procedures are inadequate, and the use of sophisticated simulation programs is often required.

(b) The system under study

The electrical system under test is illustrated in Figure 5.2. It consists of a main generator rated at 26 kV, 1200 MV A at 0.9 power factor. Power is delivered to the national high-voltage grid via two step-up transformers. A second independent source for the electrical auxiliary system comes from the 132 kV network via the start-up transformer. Power is supplied to the loads of the auxiliary system by two three-winding transformers rated at 34/17/17 MV A with a voltage ratio of 26/6.3 kV.

The electrical system of the power station is divided into one normal electrical system and one emergency electrical system.

The normal electrical system supplies loads required for start-up, normal operation and shut-down of the plant. It is divided into two identical subsystems (1N1, 1N2) for redundancy requirements. In order to avoid excessive voltage dips on the starting of large motors, separated supplies are adopted for large and small medium-voltage motors. This results in the presence of two further buses (1N3, 1N4).

The emergency electrical system supplies loads necessary for the safe shut-down of the plant in emergency conditions. The emergency system

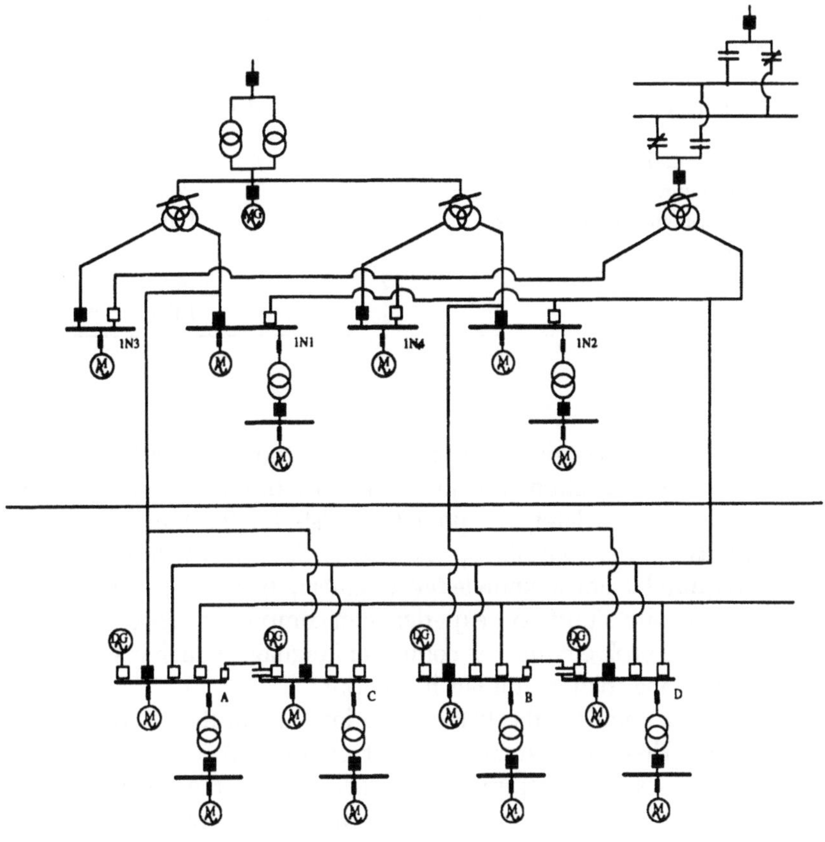

Figure 5.2 Electrical auxiliary system for a 1000 MW power plant.

is divided into two safety-related divisions (1,2) each one having two load groups (A and C for division 1, B and D for division 2). Each load group is equipped with an emergency diesel generator.

5.3.2 The IKBS approach: building the knowledge base and design procedure

The problem of designing the electrical system of a large power plant was tackled with the IKBS approach in two different ways. The first approach, which has already been mentioned [18], makes use of LISP to realize the inference engine, the design rules, and the knowledge base with all the information about the load list, equipment ratings etc. The second approach, which is described in the following, makes use of NEXPERT [29], an IKBS shell with a general-purpose inference engine and very useful editors for translating human experience into rules and objects.

Table 5.1 Some typical large electrical loads in the considered plant

Pump type	*Number*	*Pump rating (kW)*
Reactor coolant	3	6000
Emergency feedwater	2	790
Component cooling water	4	1000
Safety injection:		
low head	2	470
high head	2	1370

The two approaches give the same results, although the effort required for knowledge acquisition was quite different. In the case of NEXPERT, as will be discussed later, the potential exists for the human expert to access the computer to build up the knowledge base, with little contribution required from a knowledge engineer. In the case of LISP, an intermediary with significant experience in computer science is required to translate the problem into the computer language, which is still related more to machine than to human behaviour.

The starting point for the IKBS is the load list of the electrical auxiliary system. This load list may consist of up to 2000 electrical loads. About 100 loads are medium-voltage loads, while the remaining ones are low-voltage loads. In order to limit the complexity of the problem, the IKBS has considered in detail only the largest loads of the plant, collecting all the remaining loads into a lumped load, supplied by medium-to-low-voltage transformers. A partial list of the considered loads is given in Table 5.1.

The adopted shell could be directly interfaced with a database in which a design expert could have collected technical data concerning the electrical loads of the plant. In this prototype, the same result has been obtained by describing each load of the load list as an object, and by defining different classes of loads. For example: a general class including different types of plant loads is labelled `electrical_motor_loads`; a specific subclass, labelled `pump_loads`, is more strictly related to motor-operated pump loads. These classes have a certain number of properties, these being the attributes that the designer is expected to use during the design process. Figures 5.3a and 5.3b indicate the NEXPERT facility used to build the aforementioned classes. Each defined object (i.e. each load) will be given a number of slots (only some of which are shown in the figures) with the same attributes as its class. It is then the task of the user to give the attributes their actual value for that specific load.

(a)

Name:

electrical_motor_loads

SubClasses:

pump_loads
compressor_loads
motor_operated_valves

Properties:

busbar	(S)	MV1
description	(S)	Unknown
efficiency	(F)	Unknown
I_locked_rotor	(F)	Unknown
kVA	(F)	Unknown
kVA_short_circuit	(F)	Unknown
odd_even	(S)	Unknown
operating_conditions	(S)	Unknown
power_factor	(F)	Unknown

(b)

Name:

pump_loads

SubClasses:

Properties:

busbar	(S)	MV1
description	(S)	Unknown
efficiency	(F)	0.94
I_locked_rotor	(F)	5.5
kVA	(F)	NotKnown
kVA_short_circuit	(F)	Unknown
odd_even	(S)	Unknown
operating_conditions	(S)	Unknown
power_factor	(F)	0.87

Figure 5.3 Class editor: (a) `electrical_motor_loads`; (b) `pump_loads`.

The object editor and the object network are examples of the facilities offered by the NEXPERT shell, which are illustrated in Figure 5.4. These have been shown with reference to a specific load, namely CCWP1 (component cooling water pump number 1).

These tools are quite useful at the initial stage of design, when some parameters, such as efficiency and power factor, transformer impedances etc. are unknown but will be known later in the project development. The IKBS can pick up these parameters from similar electrical components which have been previously used, that is, from equipment performing the same function in the design of another plant or from similar component data sheets. Further information is provided to the IKBS regarding nominal currents and voltages for different types of busbars, transformers etc.

Before activating the IKBS, an initial configuration of the electrical system must be assumed. Since the aim of the proposed IKBS is to develop a one-line diagram in the same way as a human expert would, the initial configuration proposed to the IKBS is very simple. This is shown in Figure 5.5a to simply consist of a main generator which is connected to the national grid and supplies auxiliary loads of the plant by means of a single busbar MV1, with a second external supply from

```
Name:
    ccwp1

Classes:
    pump_loads
    electrical_motor_loads

SubObjects:


Properties:
    busbar                  (S) MV1
    description             (S) component_cool
    efficiency              (F) 0.94
    I_locked_rotor          (F) 5.5
    kVA                     (F) 1223.0
    kVA_short_circuit       (F) Unknown
    odd_even                (S) odd
```

(a)

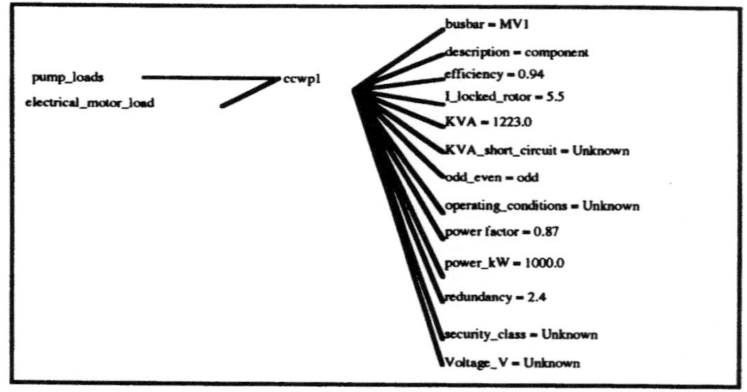

(b)

Figure 5.4 (a) Object browser for CCWP1; (b) object network for CCWP1.

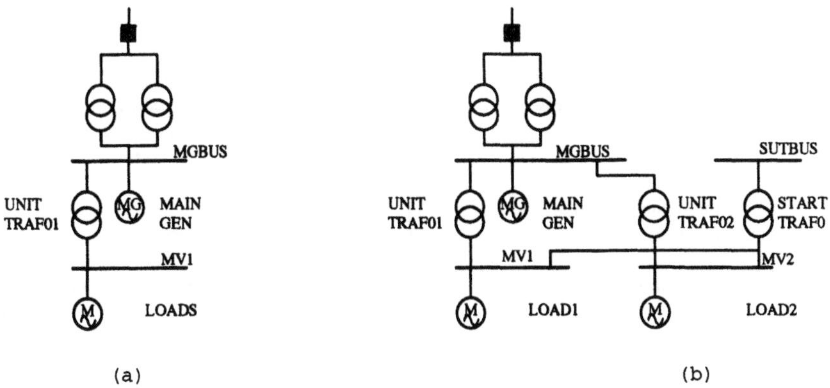

Figure 5.5 (a) Initial configuration of the expert system reasoning; (b) improvement due to the application of the first design rule.

the start-up transformer. Obviously this solution is highly inappropriate because of the loads' wide range of electrical power requirements. It is incorrect for all the loads to be supplied at the same voltage. Moreover, reliability and availability considerations require more than a single busbar supply.

The IKBS will decide upon the division into subsystems, will choose the number of buses and their voltages, will define the presence of emergency generators etc. These decisions are made on the basis of the design rules (concerning availability, redundancy, busbar rating etc.) in the knowledge base. A design constraint which must be satisfied is deduced by the IKBS when scanning the load list. The existence of redundancy among loads, indicated by a property slot, causes the IKBS to split the initial single load busbar into two buses, each of which is supplied by one unit transformer and by a second external supply. The equivalent reasoning of a human expert would have been: 'If there are redundant loads (i.e. more than one load performing the same function) then duplicate the power source in order to improve the availability of electric power, and consequently duplicate MV1 by generating the twin busbar MV2, and UNIT-TRAFO1 by generating the corresponding UNIT-TRAFO2'. The improvement in the electrical one-line diagram resulting from this first step, is shown in Figure 5.5.

An example of how the IKBS can work out this step is given in Figure 5.6, where the corresponding rule is presented in the forms used by both of the aforementioned approaches, i.e. (1) directly in LISP, and (2) in the NEXPERT environment. The effort required by the IKBS developer to build this rule into the knowledge base in the two different cases should be evident.

```
(defun dupl_source (loads)
  (cond ((not (null loads))
      (dolist (item loads)
       (if (> (load-redun item) 1)
         ; then
         (progn
          (dupl_mv1)
          (dupl_unittrafo)
          (return))
         ; else
         ; continue
         )
      )
    )
)
```

```
Description:
If there are redundant loads ie. more than
one load performing the same function,
duplicate power sources. Duplicate MV1 by
generating the twin busbar MV2 and UNIT
TRAF01 by generating UNIT TRAF02.

If
Yes    *      start.session
IsNot <|electrical_motor_load...  "1.1"
Is    <<|electrical_motor_loa...  "even"

rule 1 Actions
Create   MV2                     |busbars|
Create   UNIT_TRAF02             |unit_...
Let   <<|electrical_motor_loa... "MV2"
Let   UNIT_TRAF02.receiving      "MV2"
Let   UNIT_TRAF02.start          "mgbus"
```

(a) (b)

Figure 5.6 Example of the rule causing busbar duplication according to: (a) the LISP approach; (b) the NEXPERT shell approach.

It is worth pointing out that NEXPERT, like other shells, is capable of automatically showing a window displaying an adequate explanation of the rule under examination, with messages addressed to the IKBS user as is evident from the rule shown in Figure 5.6b.

The scanning of the load list also points out the presence of normal loads and emergency loads. A design rule in the knowledge base is fired which causes the generation of an emergency electrical subsystem. The classification of this subsystem as an emergency one necessitates the connection of an on-site supply source to the busbars. The examination of the load list also reveals that active powers required by auxiliary loads are spread in a wide range from a few kilowatts (as in the case of motor-operated valves) to thousands of kilowatts (e.g. large motors for circulating water pumps).

Different voltage levels must be used to supply such different types of loads. A simple design rule adopts standardized voltage levels for given ranges of power. The ENEL standard for power stations requires a 6 kV supply for motors rated more than 200 kW, and a 380 V supply for smaller motors. The rule used by a human expert could be expressed as follows: 'If power required by this load is over a given threshold, select a middle-voltage busbar to supply it, otherwise allocate it to a low-voltage busbar'. The IKBS version of this rule is shown in Figure 5.7 in its shell version.

```
Description:
If power required by some loads is over a given
threshold (200kW), select a middle voltage busbar to
supply them, otherwise allocate them to a low voltage
busbar.

If
>     *     <|electrical_motor_load...  200
<=          <<|electrical_motor_loa...  200

Actions
Do          MV1.voltage_V                <|elec...
Do          LV1.voltage_V                <<|ele...
```

Figure 5.7 Rule which performs load allocation to busbars according to power requirements (according to the shell approach).

The selection of the voltage level according to power requirements can also be left to IKBS users who, in response to a question posed to them by the IKBS, can suggest their own voltage level.

As no small motor was involved in the considered case, the one-line diagram resulting from the application of the rule shown in Figure 5.7 does not contain any detail about the low-voltage network and is limited to the medium-voltage electrical distribution system inside the plant.

At this stage of the design process, each load of the load list has been given a supplying busbar. Attention is also paid to attributing odd loads to odd buses and even loads to even buses in the presence of load redundancy.

Following this preliminary load allocation, some electrical calculations must be performed. The typical calculations at each busbar are power balance, verification of interrupting and peak currents, and voltage drops at steady-state and transient conditions.

Normal and emergency subsystems are considered separately as two completely independent networks. The numerical analysis is performed for different operating modes of the power plant (i.e. at different power requirements of the electrical auxiliary system). The simplified calculations are based on international standard procedures (e.g. the ANSI, IEEE and IEC Standards). The results of the computation performed for each busbar are compared with design limits which have been pre-defined (as properties of electrical equipment) in the knowledge base. Continuous current, interrupting and momentary currents, and maximum allowable voltage drop are all typical design parameters.

If a violation of these limits is detected by the IKBS, some decision must be taken. In such a situation a human expert could decide either to reinforce equipment characteristics by choosing other equipment with increased sizes from the knowledge base, or by modifying the one-line diagram structure. The IKBS only considers this second alternative and does not consider the selection of equipment different from ENEL's standards. That is, only network topological variation can be suggested by the inference engine.

Continuing with the case under study, a violation of the continuous current rating for the selected equipment at busbar MV1 causes the IKBS to split this medium-voltage busbar into two sections: one section is dedicated to the supply of very large motors (over 2000 kW), and the other section is dedicated to the remaining loads and to the medium/low-voltage transformers. The same result can be expected from the analysis of busbar MV2. Instead of duplicating each of the unit transformers supplying the initial medium-voltage busbars MV1 and MV2, a more economical solution involving two three-winding transformers (one for each bus) is suggested.

The final one-line diagram, which is shown in Figure 5.8, is a valid design. No great difference exists between the one-line diagram adopted by the human designer and the one resulting from the IKBS reasoning.

In its present state, the one-line diagram produced by the IKBS cannot be considered to be a detailed diagram ready for design application. It can, however, be successfully used for the preliminary definition of voltage levels, the internal electrical distribution and the load allocation to busbars. The adoption of graphical on-line facilities could significantly improve the IKBS's effectiveness as a powerful training tool.

Some significant advantages to be gained from the use of the IKBS presented above can be summarized as follows. The ability of IKBSs to reproduce the decision-making process of a human expert appears to be useful in answering questions regarding, for example, the number of buses and their voltage level, and the splitting of the auxiliary system electrical network into subsystems.

Solutions proposed by the IKBS could more accurately take account of the requirements of the different specialists involved in the design process. One example of a conflicting requirement is that which exists between specifying a very fast start-up for large motors (which is a typical request from the process engineer) and, at the same time, requiring only very narrow busbar voltage variations (as generally required by the electrical system engineer). The IKBS can detect this, and suggest to the user the adoption of a higher voltage level for the busbar. Consequently better coordination within the overall design could be achieved by the IKBS approach.

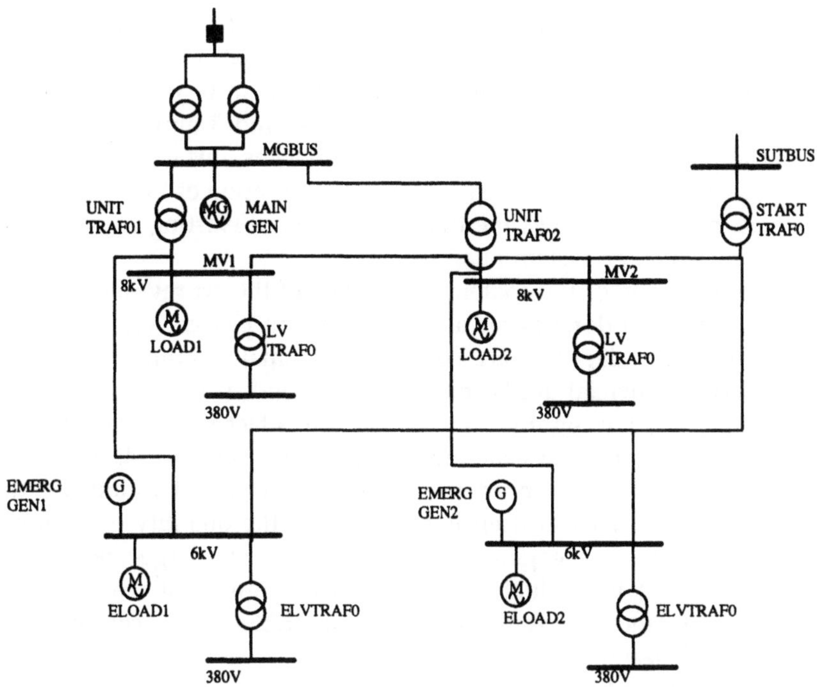

Figure 5.8 Further improvements in the design as suggested by the expert system: voltage selection and final one-line diagram.

Finally, the IKBS presented seems to be well suited to the training of new personnel involved in the design and operation of auxiliary systems.

5.4 CONCLUSIONS

In the context of this chapter, which has been devoted to the application of IKBSs to the design of electrical plants and systems, an IKBS shell was used to solve a typical design task, namely the design of the electrical auxiliary system of a power plant.

To consider this matter in more detail, some of the benefits mentioned in section 5.1.2 should be briefly reviewed. Among these benefits the following deserve mention. The decision-making capability of IKBSs allows them to solve 'if-then' questions typical of the design process. The possibility of exploring different design solutions is offered to a planner or design engineer. IKBSs also have the power to train inexperienced designers.

Future attempts to improve the efficiency of IKBSs should consider improvements to their connection to powerful power system analysis software. As this software was mostly developed in Fortran [30], communication problems between IKBSs and conventional numerical routines have been one of the most important difficulties to be overcome. As is also pointed out by Dillon and Laughton [4] and Tamura and Sasaki [31] with particular reference to planning problems, this is still an important aspect for future improvements in the application of IKBSs to power systems, because it allows the full exploitation of existing conventional software for power system analysis.

An examination of the contemporary state of the art reveals that a lot of work has been done as far as applications to power system control, monitoring and diagnosis are concerned. A smaller number of prototypes and commercial applications can be found in the area of design and planning. Some significant reasons for this fact have been reported in section 5.1.1 and are examined in detail in reference 4. The most important reason is the burden of work involved in building up the knowledge base. This in itself is the result of the quantity of information related to these engineering areas being very large and spread among different individuals [32,33]. Another significant reason is the lack of confidence in results provided by IKBSs when compared to conventional design [34].

With regard to the former observation, the tremendous rate of growth in computer technologies in recent years would suggest that much more powerful tools for knowledge acquisition will be available in the near future. It should then be easier to collect information about power system design and planning.

The latter observation is more difficult to solve. It is related to a common attitude among experienced engineers towards computer tools. Unless these tools are very user-friendly, engineers will not be inclined to use them. IKBSs can truly be the solution to this, since they are powerful aids for human–machine communication and can be used by non-specialists. Moreover, their extensive use as training tools for design and planning subjects and their constant improvement will help make engineers more familiar with information technology.

Finally, an important aspect of IKBS applications is the possibility for upgrading the design quality by systematically collecting design knowledge in terms of constraints and goals.

ACKNOWLEDGEMENTS

The author wishes to thank Professor Bruno Delfino of the Electrical Engineering Department, University of Genoa, for fruitful discussions.

Thanks are also due to Roberto Galdi and Angela Protti of the CAD Laboratory at the Electrical Engineering Department of the University of Pavia for their co-operation.

REFERENCES

1. CIGRE Working Group 38–02, Task Force No. 7 (1988) An international survey of the present status and perspective of expert systems on power system analysis and techniques, Final Report, March.
2. Hope, G.S., Zhang, Z.Z. and Malik, O.P. (1989) Expert systems in electric power systems – a bibliographical survey. *IEEE Transactions on Power Systems*, **4**(4), 1355–62.
3. CIGRE Working Group 01 of Study Committee 37 (1991) Flexibility of power systems. Principle means of achievement and approaches for facing uncertainty by planning flexible development of power systems. *Electra*, **35** (April) 75–101.
4. Dillon, T.S. and Laughton, M.A. (1990) *Expert system application in power systems* (Series in power system computation), Prentice Hall.
5. Fiksel, J. and Hayes Roth, F. (1989) Knowledge systems for planning support. *IEEE Expert*, No. 3(Fall), 16–23.
6. Bara, B.G. (1990) *Scienza cognitiva (Cognitive Science)*, Bollati Boringhieri Editrice, Torino, Italy.
7. Wenger, E. (1987) *Artificial intelligence and tutoring systems: Computational and cognitive approaches to the communication of knowledge*, Morgan Kaufman, Los Altos, CA.
8. *First Symposium on Expert System Application to Power Systems*, 22–26 August 1986, Helsinki, Finland.
9. *Second Symposium on Expert System Application to Power Systems*, 17–20 July 1989, Seattle, USA.
10. *Third Symposium on Expert System Application to Power Systems*, 1–5 April 1991, Tokyo-Kobe, Japan.
11. Mohammed, O.A. and Tomsovic, K. (eds) (1996) *Proceedings of the International Conference on Intelligent Systems Applications to Power Systems (ISAP'96)*, IEEE.
12. Hertz, A., Holen, A.T. and Rault, J.C. (eds) (1994) *Proceedings of the International Conference on Intelligent Systems Applications to Power Systems (ISAP'94)*, EC2.
13. Galliana, F. and McGillis, D. (1989) Preliminary design of a point to point EHV AC transmission system – an expert system approach, in *Seminar on 'Expert System Application in Canadian Electric Utilities'*, Canadian Electric Association, Regina, September.
14. Baggini, L., Cicoria, R., Gallanti, M., Invernizzi, A. and Pessi, E. (1991) Expert system for power system planning. SELF: an assistant to load-flow calculation, in *Third Symposium on Expert System Application to Power Systems*, 1–5 April, Tokyo-Kobe, Japan, pp. 55–62.
15. Liu, C. and Dillon, T.S. (1990) State of the art, in *Expert System Application in Power Systems*, (eds T.S. Dillon and M.A. Laughton), Prentice Hall, pp. 383–408.
16. Galliana, F.D., McGillis, D.T., Ahmad, B. and Bernard, J.P. (1989) An expert system for insulation coordination, in *Seminar on 'Expert System Application in Canadian Electric Utilities'*, Canadian Electric Association, Regina, September.
17. Puettgen, H.B. and Jansen, J.F. (1989) An expert system for the design of power plant electrical auxiliary system. *IEEE Transactions on Power Systems*, **3**(1), 254–61.
18. Delfino, B., Invernizzi, M. and Massucco, S. (1989) Artificial intelligence applications in power systems: an approach to load allocation in power plant

auxiliaries, in *International Symposium on Electrical Energy Conversion in Power Systems*, paper W-R.6, Capri, Italy, May.
19. Buiko, J. and Ajavon, A.S.A. (1989) An expert system for industrial power networks design, in *Second Symposium on Expert System Application to Power Systems*, 17–20 July, Seattle, USA, pp. 310–16.
20. Wong, K.P. and Cheung, H.N. (1987) Artificial intelligence approach to load allocation in distribution substations. *IEE Proceedings*, Part C, **134**(5), 357–65.
21. Taylor, T. and Lubkeman, D. (1989) Application of knowledge based programming to power engineering problems. *IEEE Transactions on Power Systems*, **4**(1), 345–52.
22. Assi, N.M., Kelly, J.J. and McDonald, J.R. (1991) Expert system application in electric cable design, in *UPEC – Universities Power Engineering Conference*, UK.
23. Lambert Torres, G. and Mukhedkar, D. (1989) An expert system for AC/DC substation grounding design of a power system, in *Second Symposium on Expert System Application to Power Systems*, 17–20 July, Seattle, USA.
24. Satoh, H. and Serizawa, Y. (1991) A design support system for overcurrent protection coordination on industrial power distribution systems, in *Third Symposium on Expert System Application to Power Systems*, 1–5 April, Tokyo-Kobe, Japan.
25. Kawahara, K., Sasaki, H., Kubokawa, J., Kitagawa, M. and Sugihara, H. (1991) Construction of expert system for transmission line protection, in *Third Symposium on Expert System Application to Power Systems*, 1–5 April, Tokyo-Kobe, Japan.
26. Giraldi, A. (1985) L'evoluzione del progetto unificato dei sistemi elettrici delle centrali termiche ENEL (Evolution of the standard design of the electrical systems for ENEL thermal power plants). *L'Electrotecnica*, **72**(11).
27. Waterman, D.A. (1986) *A Guide to Expert Systems*, Addison-Wesley Publishing Company.
28. Gerth, G.O. (1987) *Auxiliary System Planning*, Power plant electrical reference series, Vol. 3, EPRI EL 5036, Palo Alto, CA.
29. Neuron Data. *Nexpert Object – Fundamentals*, Neuron Data Inc., California.
30. Glavitsch, H. and Reichelt, D. (1990) A knowledge based system for security enhancement combining AI-techniques and analytic algorithms, in *Proceedings of the Tenth Power Systems Computation Conference*, Graz, Austria, 19–24 August, pp. 921–28.
31. Tamura, Y. and Sasaki, H. (1991) Expert system applied to voltage and VAR control, TF 38.06.01 – Final Report, *Electra*. No. 139 (December) 109–31.
32. Belenguer, E. *et al.* (1996) A knowledge based System for implementing insulation coordination in the power system planning process, in *Proceedings of the International Conference on Intelligent System Applications to Power Systems (ISAP'96)*, pp. 254–60.
33. Lee, S.J. and Liu, C.-C. (1996) Intelligent approach to coordination identification in distance relaying, in *Proceedings of the International Conference on Intelligent System Applications to Power Systems (ISAP'96)*, pp. 62–67.
34. Knight, V.G. (1987) Expert systems in power systems planning and operation from the viewpoint of a utility engineer, in *Proceedings of the 9th Power System Computation Conference*, Cascais, Portugal, September, pp. 687–94.

6

Intelligent knowledge based systems in condition monitoring

Barclay Weir

6.1 INTRODUCTION

This chapter will consider the uses of intelligent knowledge based systems (IKBSs) in the condition monitoring of electrical plant and systems. We shall take a broad view of what constitutes condition monitoring, and will define our terms as they arise. Firstly we shall consider different types of electrical machines and classify them according to their design and construction. This will allow us to examine and classify failure mechanisms and to describe symptoms of incipient failure. We shall then consider the process of condition monitoring by technique, measurement and data acquisition.

Having acquired data at some cost, it must then be analysed and interpreted. At this stage the data is transformed into information, and it is here that human skills are brought to bear in the analysis of the information. This analysis, when coupled with expertise, can provide a diagnosis of a problem and a valuable prognosis which may have operational (and hence economic) impact. However, these skills are themselves expensive to acquire, and are relatively rare. They tend to be specialized rather than general, and may be individualized to a high degree. The application of these skills at the point of benefit may involve geographical and other problems. For example, the expert may be distant from the problem or the problem may arise in a time scale which precludes the personal attendance of an expert at the problem site. In general, we can expect the expert to be absent for both reasons. However, when economic factors are considered, be they loss of revenue or capital investment, we recognize that lack of expertise and expert advice can prove to be costly. The case for multiple experts is readily apparent. The very nature of expertise however, means that the solution must lie elsewhere.

Given one or perhaps a few experts in a particular area of electrical or mechanical engineering (the domain of expertise), and many sites requiring the expertise, how can we provide guidance and advice as and when required? This problem has for some time been recognized as being suitable for the application of the relatively new technology of knowledge based computer systems, or more colloquially, 'expert systems'. Just as real-life expertise can cover a spectrum of knowledge and experience, we shall see various shades and colours of IKBSs. One fact will, however, be prominent in this discussion, and that is the great opportunity there exists for the application of such systems to electrical engineering and to condition monitoring of plant in particular. Englemore [1] quotes a statistic of two-and-a-half days increased uptime per year over two years for utilities using the Westinghouse diagnostic facilities at Orlando, Florida. Such a figure represents a large sum of money for the companies involved, since generating plant downtime can cost £200 000 per day.

Some of the reasons why such systems are receiving attention in relation to these fields are:

1. the important role of electrical plant in modern society;
2. the high levels of capital investment in electrical plant;
3. the increasing scarcity of engineering expertise;
4. the increasing sophistication of condition monitoring;
5. the increasing availability of computer-based engineering solutions.

6.2 TYPES OF ELECTRICAL MACHINES

Electrical energy systems encompass the generation, transmission and distribution and supply of electrical energy. For our purposes we shall include some of the uses of electrical energy, in particular large induction motors used for transforming electrical energy into mechanical power. We include these machines since they play an important role in generating stations.

A modern industrialized society is completely dependent on these systems as the most efficient means of providing energy at its point of use. The pervasive nature of these systems is evident in the unconscious acceptance by users of the instant and continuous availability of electrical energy both in the home and in the workplace. We can summarize the objective of such systems as:

1. the generation of electrical energy in sufficient quantity, at the most suitable location;
2. its transmission in bulk to load centres;

3. its distribution and supply to individual customers in a suitable form and quality at lowest economic price.

The importance to modern societies of such systems cannot be overstated, and henceforth will be assumed as axiomatic. However, the breadth and depth of engineering and scientific applications involved in the operation of these systems are largely unappreciated by both public and government in such societies. We can classify the component parts of such systems as follows:

1. network – covering transmission, distribution, supply;
2. static plant – including transformers, switches, automatic voltage regulators (AVRs), circuit breakers and static VAR compensators (SVCs);
3. rotating plant – including motors and generators.

Generally speaking, the failure of a network or part of a network is caused by the failure of static plant or rotating plant associated with the network. This allows us to consider failure mechanisms in static and rotating plant as primary causes of disruptions in energy systems. There are other primary causes of failure, such as human error, weather etc., but these are essentially random and give rise to the same effects, and so are subsumed in the following discussion.

Failures in static plant can be categorized as:

1. electrical – due to electrical effects alone, e.g. short circuits, open circuits;
2. electromechanical – due to mechanical effects, e.g. vibration, magnetostriction.

Similarly, failures in rotating plant can be categorized as:

1. electrical – due to electrical effects alone, e.g. short circuits, open circuits;
2. mechanical – due to mechanical effects, e.g. vibration, excessive load, shock load.

In practice, nearly all failures in operating plant are due to some mechanical root cause.

This chapter will focus on rotating plant – turbines, generators and induction motors – and on power transformers as exemplars of plant. In particular we will consider IKBSs applied to the condition monitoring of plant, which, as it turns out, covers a wide range of systems. We should not be surprised by the scope, in view of the application. We must however, restrict our considerations to cover the generic aspects of such systems in order to highlight the principles being applied.

6.3 CONDITION MONITORING

Condition monitoring can be taken to mean the continuous or regular acquisition of plant parameters. The simultaneous or subsequent evaluation of these measurements provides a global indication of the health of the plant or equipment being monitored. The purpose of condition monitoring is implicit in this description, since it would not be sensible to invest in such systems without some aim in mind. The aim is to provide some pre-emptive knowledge of faults or condition relating to the plant. In turn, this knowledge may be used to economic advantage, e.g. a machine may be shut down for maintenance prior to damage or a motor may be retained in duty for a specific time period secure in the knowledge that it will not sustain irreversible damage. We see here a distinction between condition monitoring and plant protection, the latter being designed to remove the shut-down decision from the operator and conservatively prevent plant damage. Protection systems can, however, operate at a fairly crude economic level, for it may be that continued operation of a plant item makes more economic sense in a particular circumstance than shutting down and losing revenue. For instance, the plant protection may be operating on erroneous data, or the equipment trip settings may be extremely conservative.

Another aspect of condition monitoring is its impact on plant maintenance regimes. By investment in such systems, operators may feel able to reschedule planned maintenance outages. By increasing the fixed time intervals, operators can, as a result, benefit by avoiding unnecessary component replacement and from increased plant availability.

We can define condition monitoring as the process of (1) acquisition, (2) analysis and (3) interpretation of signals from sensors installed in or applied to operational plant.

1. Data acquisition is performed in a number of different ways. It may be simply logged by plant attendants taking readings from gauge boards. Alternatively, it may be automatically recorded on strip charts, or stored as history files in computer memory or disc storage. Each method depends on the primary transducers being available for the measurements to be made. In practice, all three methods may be in use at any one site for different purposes. We shall henceforth restrict our discussion to data acquired by computer.
2. Analysis of the data acquired from plant is the first step in its conversion into knowledge about the condition of the plant. For example, data relating to bearing vibration at, say, 20 bearings of a large steam turbine/alternator set represents 20 measurements. All readings may be within their respective alarm levels, but an abnormally low reading at one bearing may mean the bearing is underloaded or that the transducer is faulty, or even that the recording device is faulty. Experienced

operators use the implied relationships between bearings to decide which case is true in a particular circumstance. This skill is applied daily in many plants where complex equipment is in operation. In arriving at a judgement, the operator does not consider each of the indications as individuals, but as parts of an overall vibration 'picture' of the machine.

3. Where the plant indications are giving rise to concern, interpretation of the analysed output is needed to establish what course of action (if any) should be taken. The interpretation requires a high degree of skill and is dependent on deep knowledge (analytic knowledge) coupled with experience of the monitored plant.

Monitoring can be a continuous process of data acquisition or a sampling procedure whereby a series of 'snapshots' of operational plant parameters are acquired. In both cases the parameters are usually specified by the manufacturer. These are normally indicative of the 'health' of the plant in a general sense. It is normal for these parameters to be continuously or regularly indicated and recorded. Instrumentation such as multipoint sampling chart recorders is still in use for this duty. Increasingly however, computer-based systems are being introduced. These can be integrated with control systems, and used to drive visual display systems which provide plant mimic diagrams with current parameter values being updated at regular time intervals.

For electrical machines, it is not always possible to perform direct measurement of indicative parameters. This may be due to inaccessibility or transducer unavailability. For example, direct measurement of partial discharge activity in generator stator windings is difficult to achieve, and novel transducers are actively being developed for duty outside the stator shell. The health of transformer windings is normally inferred from indirect measurements of related parameters for similar reasons.

Supply waveforms are usually more accessible in machines like induction motors, and the information content of these waveforms is high. Strong inference may be derived from these measurements by standard techniques such as Fourier analysis, and we shall return to this technique when considering IKBSs applied to large induction motors.

Mechanical symptoms are also used to monitor the health of plant. In particular, the vibration of rotating machines is widely used for on-line health indication. It is also used for the off-line analysis of plant status, since vibration is both a general indicator of whole machine health, and a specific symptom of local (to the transducer) problems. Such problems may be very localized and may themselves be the result of damage or problems not directly measurable.

A vibration 'signature' or 'fingerprint' is often recorded when plant is returned to service following overhaul. This fingerprint is then used as a datum against which subsequent operational phases are measured. The amount of information in vibration signals is very high, and the skilled interpretation of these signals can lead to the recognition of several classes of problems associated with rotating machinery.

For static plant, vibration in the form of magnetostriction can be used to assess general health. The chemical analysis of gases in insulating oil is used to provide a measure of the health of transformer windings by an indirect means. Similarly, analysing the coolant gas used in generators can provide an early warning of insulation breakdown.

All of the foregoing comments assume that suitable transducers are available, and that data acquisition equipment provides the raw data for skilled interpretation. It is usual for manufacturers of high-capital plant to provide a basic set of sensors with the machine. However, these are aimed at operations, and may not be suitable for diagnostic purposes. As technology progresses and transducer and other equipment costs fall, there is a trend to retrofit additional sensor sets to plant. These additional transducers provide both new measurements and different views of parameters already instrumented. Where retrofitting of fixed instrumentation cannot be justified in cost terms, the use of portable test equipment may be applicable.

In all of the above cases – permanent, retrofit permanent and portable – there is a problem associated with the sheer volume of raw data. Given that the vast majority of plant will be healthy at any given moment in time, how can all this data be scanned for early indications of problems? The interpretation of these measurements calls for skill and expertise. The most basic technique, that of visual survey, is the one most difficult to mechanize. An expert interpreter can survey chart recordings of bearing vibration and quickly decide if there are any signs of problems which require a more detailed analysis. Similarly, a less-skilled individual can compare sets of recent measurements against post-overhaul fingerprint data, and decide whether a machine is under stress or not. It is normal for operators of high-capital plant to maintain historical information of operating parameters for the life of that plant. These records can be used to recognize long-term drift or gradually worsening conditions caused by some underlying problem associated with operating conditions.

There are other methods of assessing the health of plant. These are associated with a more sophisticated analysis of the fundamental data acquired from suitable transducers. For example, spectral analysis of supply-current waveforms can be used to derive a figure of merit for the internal state of rotors in large induction motors. Again, this analysis requires great skill and deep understanding of the principles underlying the technology and engineering associated with this class of machine.

The benefits of computerized monitoring systems are manifold. Among the more obvious benefits are:

1. Rapid display of information to operators;
2. Formatted information which aids interpretation;
3. Elimination of hand logging of parameters;
4. Improved accuracy of logged parameters;
5. Facilities for review of historical data.

These aspects are explored in depth in Hayashida [2].

The integration of well-designed graphical interfaces with computer-based condition-monitoring systems provides valuable decision support for plant operators. These interfaces should be based on accepted standards, and use colour and high resolution displays. Gray [3] claims an increase in generator availability from 95% to 100% over two years of operation when a diagnostic user interface was integrated with an already existing IKBS. The system was delivered to a nuclear plant in the USA, and linked via computer communications to the diagnostic centre in a remote location.

From the above, we can summarize condition monitoring as covering the acquisition and interpretation of data related to the general health and particular problems associated with static and rotating plant. Let us proceed now to consider in more detail how IKBSs can be brought to bear on the interpretation phase, and thereby provide cost effective expertise at the point of need when required.

6.4 IKBSs IN CONDITION MONITORING

knowledge based techniques encompass a range of technologies. We are concerned with the delivery of expertise in a suitable form and in a usable fashion. In practice, this means computer-based systems. The reduction in costs associated with computer equipment has been marked in recent years. This has allowed technologies which were perceived as not being cost effective just a few years ago to emerge from the computer science laboratory and be applied in industry. IKBSs have traditionally been perceived as esoteric, expensive and risky to apply. These perceptions have undergone rapid change in the last few years, and IKBSs are now emerging in the workplace.

It would not be sensible in the majority of cases to introduce computer-based IKBSs to a working environment where computers were not already in use. The unnecessary burden this would place on operational engineers would be unacceptable. Fortunately, in the electrical generation and supply industry this situation is unlikely to occur. Computers are now an integral part of daily operations, and are applied in control rooms throughout the industry. Presently they are used in roles ranging from system monitoring (SCADA), through off-line condition monitoring, to maintenance planning. It is fortunate, therefore, that computer-based IKBSs have a ready-made potential user population throughout the industry. The form which any particular delivery will take is dictated by a number of factors, such as:

1. company policy
2. systems already in existence
3. cost factors
4. potential economic benefit
5. perceived need.

Assuming the existence of suitable transducers or signals available from fixed or portable sensors, we can consider the case for IKBSs as having been made. The feasibility, or otherwise, of designing a successful IKBS and the probability of delivering a useful program to cost are still to be considered [4]. We next consider potential and actual areas of application.

6.5 APPLICATIONS

The tendency toward increasing the size of turbine generators and the amount of system control vested in these large machines means that downtime of just one machine can invoke substantial economic penalties in lost revenue and replacement generation charges. Large steam turbines operating at high temperature and pressure represent considerable capital investment for a utility company. These machines are engineered to surprisingly fine tolerances, and operate over large temperature ranges. They are subject to a number of mechanisms which can impair efficient operation or even cause damage resulting in automatic tripping followed by investigative stripdown. Similarly, the modern generator with hydrogen and/or water cooling operates with extremely small clearances [5,6]. These generators are the result of complex design decisions constrained by many factors such as number of poles, system operating frequency etc.

The equipment associated with these generators include pilot and main exciters, and AVRs. Turbines have speed governors, which are themselves very complex pieces of equipment, forming a domain of expertise in their own right.

Large power transformers are fundamental to the operation of transmission systems, and are extremely costly items to replace. In the main, these items are instrumented to only a rudimentary level. This is due to their longevity, many of them pre-dating modern instrumentation techniques. Usually protection schemes are relied upon to provide safeguards for these expensive items of equipment.

Transmission systems are protected by complex protection and signalling equipment. As mentioned earlier, this equipment itself is subject to faults. Understanding which combination of equipment faults can expose the system to an unacceptable risk of damage following system faults, and which combinations are acceptable, is a scarce expertise among system operations engineers. Yet this expertise can eliminate unnecessary switching operations which would reconfigure a system which is still adequately protected by alternative signalling routes.

Large cage-rotor induction motors are the workhorses of industry, and are employed widely within the power industry for duties such as fan drives, pump drives etc. These motors vary in power output size, and in physical size. Again, they represent considerable capital investment, and can be subjected to rigorous duty cycles. Fault mechanisms within these machines are progressive in nature, and a machine can withstand its duty cycle almost up to the time of breakdown. Repair or replacement of a machine which has suffered complete breakdown can be a costly affair. If early or timeous warning of incipient failure can be made, the motor can be replaced or planned for maintenance, and a cheaper repair effected [7,8]. It may be possible for a motor to be maintained at duty under a more rigorous monitoring regime, until a convenient time for its repair can be planned.

6.6 SOME SUCCESSFUL APPLICATIONS

Two of the most successful IKBSs to have been applied in the power industry are used for diagnostic purposes related to turbine generators. The first of these – TURBOMAC – is employed by The Hartford Steam Boiler Inspection and Insurance Company in the USA. The second to be considered is the on-line diagnostic facility GenAID used by Westinghouse in Florida. Each of these systems is very large, and is accessed on a payment basis by client companies. In the same vein, GEMS (Generator Expert Monitoring System) developed by the Electrical Power Research Institute (EPRI) in the USA has been employed by Ontario Hydro to provide decision support facilities for generator plant operators. Hartford Steam and Boiler also deploy an IKBS named TOGA (Transformer On-line Gas Analysis) in relation to large power transformer condition monitoring. Both TURBOMAC and TOGA were constructed using rule induction, and are considered to be very successful IKBSs. It is worth noting that TOGA is employed to assist, rather than replace, an expert in the analysis of gas chromatography results. These systems will be looked at in some detail in the following sections.

Lastly, we will consider a software package named MOTORMON, from the well-known condition-monitoring software company Entek.

This is not marketed as an IKBS *per se,* but we will see that expertise is designed into the system at a fundamental level.

6.6.1 TURBOMAC

This sophisticated knowledge base deals with vibration symptoms in turbines, and is used by more than 60 utility and petrochemical companies in the USA via computer networks [9,10]. It contains nearly 10 000 rules represented in just under 2 MB of compiled code. The system has been available for diagnostic consultations since February 1988. The design of the facility was undertaken in 1984, using human experts and with reference to the Sohre chart [11]. A rule induction tool called RuleMaster [12] was used to induce rules from examples. The system is implemented as a data-driven, forward-chaining, consultative IKBS. It represents 139 queries, 9700 rules and some 40 possible diagnoses. Details of the rule induction can be found in references 13 and 14. TURBOMAC illustrates the case history approach to finding underlying rules which may be difficult to express verbally, but which can be employed to economic advantage once symbolized in a computer knowledge base.

TURBOMAC was developed with the aid of the Sohre charts as mentioned above. These describe 31 basic causes of turbine vibration, in 11 categories relating to 88 symptoms which are presented in eight groups. The charts provide detailed information in a fashion which is easily used by a rule induction tool to produce a 'first cut' knowledge base. TURBOMAC was produced in the incredibly short time of four weeks. This timescale was due to both the existence of the Sohre charts, and the employment of rule induction software. Following the creation of the knowledge base, a user interface still had to be designed and implemented.

The deployment of TURBOMAC is via a consultative session. Users dial up from the problem site into a centralized computer facility, and undertake a question and answer session. It should be noted that users do not see the actual rules. The appearance of the user interface is reproduced in reference 9. Each question presented to the user has four possible answers: 'yes', 'no', 'maybe' and 'don't know'. By weighting these responses according to the answers received and keeping a positive and negative score for each of the possible 139 diagnoses, the system arrives at most likely and likely diagnoses.

6.6.2 GenAID

GenAID is a generator diagnostic system offered as a centralized facility by Westinghouse in Orlando, Florida. Recognizing that a significant proportion of the generating capability of a utility company is needed just to provide for generator unreliability, gives a strong case for the development of methods to obtain better generator performance. This, combined with the high capital cost of standby plant, indicates the value of applying any technology which can reduce this financial burden on a company. A typical mix of monitoring equipment on a large generator would indicate the following:

1. core monitor
2. stator-winding temperatures
3. cooling-system temperatures
4. core temperatures
5. core vibration
6. frame vibration
7. end-winding vibration
8. bearing vibration
9. cooling-water parameters
10. seal-oil temperature
11. excitation parameters.

Westinghouse started a development programme in 1976 aimed at providing a computer-based diagnostic capability for turbine generators. In 1982, work was started on an IKBS approach to the same problem domain. This work has resulted in a very successful data-driven IKBS which operates on sensor output data. The program was written in LISP and runs on VAX hardware under VMS [15,16]. The system is centralized, like TURBOMAC above, and has a number of features which are likely to prove fundamental in future developments of similar systems [17].

Initially the rule base was composed of five subunits:

1. stator winding and magnetic core
2. hydrogen auxiliary system
3. seal-oil auxiliary system
4. generator field and excitation system
5. miscellaneous systems.

Classical knowledge-engineering techniques were employed. Interviews with numerous experts were carried out over many months. These were conducted by a qualified power engineer who was also an expert in computer system design and in artificial intelligence. This resulted in the identification of 900 rules relating to 185 conditions. The input data was acquired from 102 on-line transducers. The resulting on-line monitoring system is very different from the consultative system described above. Any on-line continuous monitoring system is potentially more powerful (in application terms) than a consultative system: it can diagnose fault conditions as they arise or develop, thus allowing for corrective action to be taken or planned.

6.6.3 GEMS

Since the mid-1980s the Electrical Power Research Institute (EPRI) in the USA have expressed a strong interest in developing IKBS applications for power plants. One application to arise from this research is GEMS (Generator Expert Monitoring System) developed jointly between EPRI and Ontario Hydro. GEMS has been applied on-line at the Nanticoke Station of Ontario Hydro and the Oswego station of the Niagara Mohawk Power Corporation since 1990.

Variations in sensor readings may indicate either changing plant behaviour or a sensor malfunction. Thus, close monitoring of changes in sensor readings might provide early indication of plant disturbances. In addition, incipient faults could be captured by consideration of variations in measured plant parameter trends.

GEMS [18, 19] contains an IKBS module and a status evaluation process. The role of the status evaluation process is to assess changes in sensor readings. Acquired sensor readings and their trends are categorized into ranges which have a related problem belief probability. The IKBS applies the information derived by the status evaluation process to ascertain the existence of problems believed to exist on the machine. In addition, for each diagnosed problem, a confidence measure between one and zero is derived.

In a complex engineering system, fault behaviour can take effect over a range of timescales. Moreover, it is often the case that minor disturbances can gradually become severe problems if left unattended. Thus, GEMS augments its basic diagnostic information with urgency and severity factors, defined between one and zero. The urgency factor gives an indication of the urgency of the problem based on the time estimated for sensors to reach their defined critical limits. For example, an urgency factor of one indicates a problem requiring immediate attention, possibly a generator shutdown. The severity factor indicates the effect of the diagnosed problem on the integrity of the generator. This is given in discrete levels, as follows.

1.0 – problem could result in an extended generator outage.
0.8 – problem could result in damage to the generator.
0.6 – problem could result in a de-rating of the generator.
0.4 – problem could result in a partial loss of generator life.
0.2 – problem will not adversely affect the generator.

Given that potentially serious problems may require urgent attention, GEMS helpfully supplies recommendations of diagnostic actions that could be taken to confirm the problem. However, given that even such checks might be of limited use under severe time constraints, GEMS additionally supplies recommendations for immediate corrective action for diagnosed problems.

A further feature of GEMS is the Installation Advisor program which permits tailoring of the GEMS software to a specific implementation site. A significant amount of the knowledge used by GEMS is thus gathered through the use of the Installation Advisor. This includes information relating to: type and location of sensors; algorithms for calculating sensor ranges; operator advice messages; machine design characteristics and modelling information. The Installation Advisor supports this process through a table-driven user interface.

6.6.4 TOGA

The most successful IKBS to deal with the diagnosis of large transformers is from the same stable as TURBOMAC above. It has been developed by the same company, Hartford Steam Boiler Inspection and Insurance Company, and used the same tool (RuleMaster) for rule induction. TOGA makes use of the data resulting from the chromatographic analysis of the gas products formed when oil or cellulose breaks down in an oil-cooled transformer. The system is used to help the working expert in a diagnosis, and consists of some 250 rules. A 3:1 cost benefit for the insurers is claimed, i.e. $3 saved for every $1 spent [20,21].

The failure of a large power transformer would be a significant cost for a utility company to bear. Transformers represent major elements in generation and transmission systems. Failure modes are usually sudden, and it is difficult to discern evidence of incipient failure. Rates of failure are low (of the order of 2% per annum), but service disruption and forced customer outages which follow a failure can incur large financial penalties. It is standard practice for companies to possess spare transformers and to have installed redundancy in sensitive areas. This represents non-working capital, and any means of reducing this burden is obviously desirable.

Despite the absence of intricate mechanisms, and the small number of component parts, the power transformer is nevertheless a complicated system. Moreover it is difficult to ascertain its health by external observation alone. It is necessary, therefore, to monitor some internal parameters. The major subsystems of a transformer are:

1. windings
2. insulation
3. oil
4. core laminations.

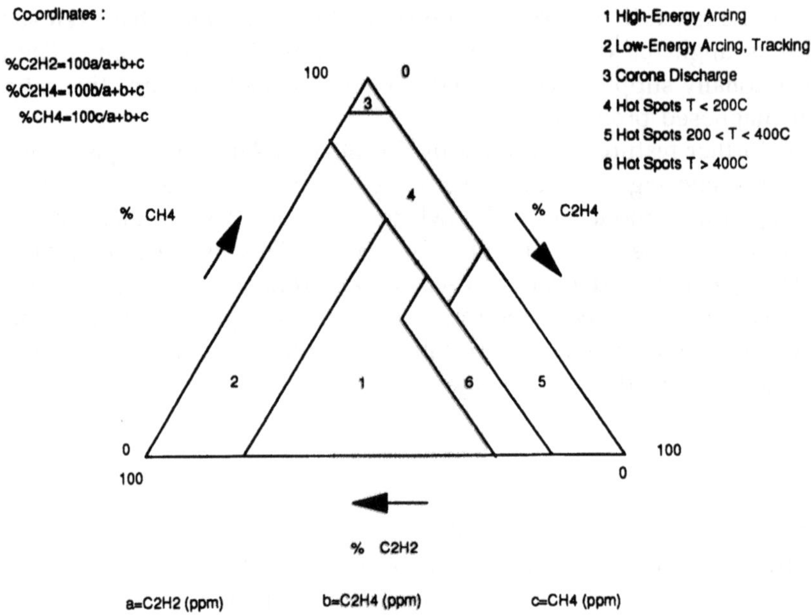

Figure 6.1 Duval's triangle.

Data acquisition from internal sources is difficult under normal operating conditions, but cost-benefit analysis shows that an IKBS which can detect incipient failure can be extremely beneficial.

Crowley [21] presents relationships between eight failure modes and seven observable quantities. Of these observables, gas-in-oil is evident in six of the eight failure modes. Marin [13] describes the maintenance and troubleshooting of transformers where the relationships between gassing and six possible diagnoses are given in the form of Duval's triangle, here reproduced as Figure 6.1.

An implementation of this diagnostic in rule form follows:

```
" if %C2H2 < 10, then zone_345, else zone_126 "
" if zone_345 and (%CH4 > 95 and %C2H4 < 5) and
  %C2H2 < 5,
        then zone_3,
        else if (%C2H2 > 50 and %CH4 < 50
                  then zone_5,
                  else zone_4 "
" if zone_126 and %CH4 < 85 and %C2H4 < 25,
        then zone_2,
        else if %CH4 < 45 and %C2H2 < 25 and %C2H4 >
  40,
```

```
                then zone_6,
                else zone_1 "
```

The diagnoses listed above are identified in Table 6.1.

In TOGA, the diagnosis of corona (zone 3 above) is represented by a rule in the form:

```
" if cor_ratio is > 4:
        if temperature is low:
              if H2_level is low:
                    then corona is unlikely
              if H2_level is medium or high
                    then corona is likely
        if temperature is moderate:
              if H2_level is low:
                    then corona is unlikely
              if the H2_level is medium:
                    if thermally generated gases are
                    absent then corona is possible
                    if thermally generated gases are
                    slight or present
                    then corona is unlikely
        else corona is possible
        else corona is unlikely
        else corona is unlikely "
```

where `cor_ratio` represents the correlation between hydrogen `H2` and acetylene `C2H2`, and the symbol '`:`' can be read as 'and'.

This rule was induced by RuleMaster from the expert example shown in Table 6.2. There is much to be gained from studying Table 6.2 as a vehicle for conveying the basic information relating to corona discharge, and the evidence which appears in gassing. The induced rule is very complex, and it should be noted that the phrase `else nil`, which would be present if the rule were cast by hand, is suppressed in this

Table 6.1 Identification of power transformer diagnoses

Zone	*Diagnosis*	*Notes*[a]
1	High-energy arcing	$I > 20I_n$
2	Low-energy arcing	–
3	Corona discharge	–
4	Hot spot	$T < 200°C$
5	Hot spot	$200 < T < 400°C$
6	Hot spot	$T > 400°C$

[a] T = temperature; I = measured current; I_n = normal measured current.

Table 6.2 An expert example relating to corona in a power transformer[a]

H_2	*Thermal*	*Correlation*	*Temperature*	*Corona*
High	–	>4	Moderate	Possible
Medium	Absent	>4	Moderate	Possible
High	–	>4	High	Possible
Medium	Absent	>4	High	Unlikely
High	–	>4	Low	Likely
Medium	Absent	>4	Low	Likely
Medium	Present	–	Moderate	Unlikely
Medium	Slight	–	Moderate	Unlikely
Low	–	–	–	Unlikely
–	–	<4	–	Unlikely

[a] '–' represents 'don't care', so the last line in the table should be read as 'If the correlation ratio of hydrogen and acetylene content is less than 4, then corona discharge is unlikely'.

representation. As an exercise the reader should rewrite the rule inserting 'ands' and `else nil` where appropriate. The whole process described above, which leads to the creation of just one rule in a system having more than 200 rules (some of which will be more complex, some less complex), gives an idea of how much expertise can be present even in a restricted domain like transformers.

TOGA does not replace human expertise, rather it is used to screen the results of laboratory analyses of oil samples. It is used in conjunction with a database of previous results for known transformers, and provides a tracking system for the health of these transformers. The results are inspected by a human expert who can over-ride the diagnosis produced by TOGA. One of the interesting aspects of this system is the relationship between the laboratory analysis results and the IKBS. TOGA was built using the results available from one particular laboratory and interpreted by one expert. Howes [22] explains that this aspect means that customers who wish to use the system but who provide their own chromatography results must be suitably cautious about the diagnosis until TOGA can be 'calibrated' with results originating in other laboratories. For example, note phrases like 'high' and 'slight' in Table 6.2.

6.6.4 MOTORMON

MOTORMON from the condition monitoring software company Entek is not marketed as an IKBS, but in fact forms an interesting example of a diagnostic system coded in a 'traditional' computer language – Compiled Basic in this case. It can, nevertheless, be considered to be an IKBS since it incorporates a large amount of practical knowledge gained over a period of time. This experience was all related to the diagnosis of problems associated with cage rotor induction motors [23]. This expertise was allied to a body of 'deep knowledge' (an analytic mathematical model) of the principles of induction motors. The resulting diagnostic program was 'productized' by the software company, and marketed successfully on a world-wide basis. The ongoing development of MOTORMON has resulted in a very successful system which has been sold into a broad spectrum of markets including:

1. oil production and refining
2. gas terminals
3. generating stations
4. water and sewerage
5. shipping
6. steel
7. paper mills.

The present status of the system as a mature software product is the result of a careful development over the past decade, by a partnership of academic research and industrial feedback. The timescales for its development are:

1. 1981–83, development of fundamental theory;
2. 1984–85, field trials and knowledge acquisition;
3. 1986–89, resolution of novel situations;
4. 1989–91, design and implementation of load influence rules.

Presently, an expert front end is in preparation, which will guide the user toward a secure assessment of rotor bar damage.

The system uses a spectrum analyser to extract measurements from the supply waveform at the motor terminals. This can be carried out *in situ* using appropriate transducers, or by laboratory analysis of analogue recordings of the waveform. It is well known that broken bars can modulate the supply waveform, and the relationship of sideband energy content with that of the fundamental supply frequency can be derived fairly readily. What MOTORMON does is to relate these measurements to real-world observations and experience, which have been correlated with confirmation derived from stripdown examinations carried out on actual cases. It is this feedback which has been designed into the diagnostic software, together with the validation of the diagnoses, which has made this system so successful.

6.7 SUMMARY

The case for IKBSs in general within the electrical supply industry is easily made. For condition monitoring of plant in particular, a number of other factors must be taken into consideration:

1. the existence or otherwise of appropriate plant transducers;
2. the existence or otherwise of computer-based data acquisition;
3. the cost of retrofitting transducers;
4. the cost of implementing a computer-based data acquisition system;
5. the capital cost of plant to be covered.

Assuming that the case for an IKBS has been made, the decision to purchase an off-the-shelf or bespoke system must be taken. There is a spectrum of costs associated with this decision, and there are complicating factors, such as:

1. computer sizing;
2. on-site delivery or centralized via computer communications;
3. availability of domain expertise;
4. availability of computer expertise.

It may be that the purchasing of dial-up rights into a commercially available system makes better economic sense.

Lastly, there are the questions of company culture and management structures. A particular company may embrace new technologies more slowly than another. Also, a company may have a strongly autonomous site management policy, whilst another may be heavily centralized.

All of the above can impact to a different degree on the fundamental decision to deploy IKBSs in the condition monitoring field. In all of this one thing is certain – the world market within the energy supply business for such systems is very large. The potential savings in economic terms are equally large. These savings arise at almost every level of operations within an energy company. For these reasons, if no others, it is certain that IKBSs will play an increasingly important role in the operations of energy supply companies. This accounts for the significant research and development of new IKBS-based condition-monitoring systems [24–28]. As the costs associated with computer hardware and software continue to decline, and the costs of engineering expertise continue to escalate, these systems will increasingly offer an attractive and cost-beneficial solution to the problem of scarce resources in the increasingly complex, technology-based, energy supply industry.

REFERENCES

1. Englemore, R.S. (1990) Expert systems: a glimpse into the 1990's, in *Expert Systems Applications for the Electric Power Industry* (ed. J.A. Naser), Hemisphere Publishing Corporation, vol. 1, pp. 1–11.
2. Hayashida, R.D. *et al.* (1991) Turbomachinery monitoring systems capture and analyze vibration data. *IEEE Computer Applications in Power*, July, 38–43.
3. Gray, R.F. and King, I.J. (1991) User interface graphically improves generator AI diagnostics. *IEEE Computer Applications in Power*, July, 16–20.
4. Crowley, T.H. *et al.* (1990) Expert system for on-line monitoring of large power transformers, in *Expert Systems Applications for the Electric Power Industry* (ed. J.A. Naser), Hemisphere Publishing Corporation, vol. 1, pp. 629–60.
5. Harper, K.E. *et al.* (1990) Lessons in deployment of successful on-line expert diagnostic systems, in *Expert Systems Applications for the Electric Power Industry* (ed. J.A. Naser), Hemisphere Publishing Corporation, vol. 1, pp. 181–99.
6. Lloyd, B. *et al.* (1990) A generator expert monitoring system, in *Expert Systems Applications for the Electric Power Industry* (ed. J.A. Naser), Hemisphere Publishing Corporation, vol. 1, pp. 611–21.
7. Thomson, W.T. *et al.* (1986) On-line diagnostics of large induction motors, NATO ARW, University of Leuven, August.
8. Thomson, W.T. and Rankin, D. Case histories of on-line rotor cage fault diagnosis, Department of Electrical Engineering, Robert Gordon's Institute of Technology, Aberdeen, UK.
9. Klimczac, B. (1990) TURBOMAC: Networked delivery of problem solving knowledge, in *Expert Systems Applications for the Electric Power Industry* (ed. J.A. Naser), Hemisphere Publishing Corporation, vol. 2, pp. 807–23.
10. Stuart, J.D. and Vinson, J.W. (1985) TURBOMAC: an expert system to aid in the diagnosis of causes of vibration-producing problems in large turbomachinery, in *Proceedings ASME International Computers in Engineering Conference*, American Society of Mechanical Engineers New York, vol. II, pp. 319–28.
11. Sohre, J.S. Turbomachinery problems and their correction, in *Sawyer's Turbomachinery Maintenance Handbook*, vol. II, Norwich, CT.
12. Michie, D. *et al.* (1984) RuleMaster: a second generation knowledge engineering facility, in *Proceedings of the First Conference on Artificial Intelligence Applications*, Denver, CO, December, pp. 591–97.
13. Marin, M.A. *et al.* (1990) Co-operating expert systems for diagnosis of electrical apparatus, in *Expert Systems Applications for the Electric Power Industry* (ed. J.A. Naser), Hemisphere Publishing Corporation, vol. 1, pp. 623–38.
14. Riese, C.E. and Stuart, J.D. A knowledge engineering facility for building scientific expert systems, Radian Corporation, 8501 MoPac Blvd., PO Box 9948, Austin TX.
15. Gonzales, A.J. *et al.* (1986) On-line diagnosis of turbine-generators using artificial intelligence, *IEEE Transactions on Energy Conversion (USA)*, **EC-1**(2), 68–74.
16. Osborne, R.L. *et al.* (1985) Increased power plant availability through on-line diagnostics based on artificial intelligence, in *Proceedings American Power Conference*, Chicago, April, pp. 539–43.
17. Harper, K.E. *et al.* (1990) Lessons in deployment of successful on-line expert diagnostic systems, in *Expert Systems Applications for the Electric Power Industry* (ed. J.A. Naser), Hemisphere Publishing Corporation, vol. 1, pp. 181–99.

18. Lloyd, B., Park, W. and White, J. (1989) Generator expert monitoring system, in *Proceedings of the American Power Conference*, pp. 393–99.
19. Divakaurni, S.M. and Scheibel, J.R. (1992) Development and testing of turbine and generator expert monitoring systems, in *Industrial Applications Of Knowledge Based Diagnosis* (eds C. Guida and A. Stefani), Elsevier, pp. 51–76.
20. Riese, C.E. (1990) A knowledge engineering facility for building scientific expert systems, in *Expert Systems Applications for the Electric Power Industry* (ed. J.A. Naser), Hemisphere Publishing Corporation.
21. Crowley, T.H. *et al.* (1990) Expert system for on-line monitoring of large power transformers, in *Expert Systems Applications for the Electric Power Industry* (ed. J.A. Naser), Hemisphere Publishing Corporation, vol. 1, pp. 639–660.
22. Howes, J.R. (1990) TOGA: the evolution of an expert system, in *Expert Systems Applications for the Electric Power Industry* (ed. J.A. Naser), Hemisphere Publishing Corporation, vol. 1, pp. 661–78.
23. Leith, D. Expert systems in A.C. induction motor fault diagnosis, Ph.D Thesis, Department of Electrical Engineering, Robert Gordon's Institute of Technology, Aberdeen, UK.
24. Kezunovic, M. *et al.* (1996) Implementation framework for an expert system for generator monitoring, in *Proceeedings of the International Conference on Intelligent System Applications to Power Systems (ISAP'96)*, pp. 140–44.
25. McDonald, J.R., Moyes, A. and Georgin, A. (1996) Turbine generator plant monitoring: towards integrated systems, in *Proceeedings of the International Conference on Intelligent System Applications to Power Systems (ISAP'96)*, pp. 151–55.
26. Tomsovic, K. and Baer, B. (1996) Methods of approximate reasoning for power system equipment condition and reliability analysis, in *Proceeedings of the International Conference on Intelligent System Applications to Power Systems (ISAP'96)*, pp. 310–15.
27. Georgin, E., Bordin, F., McDonald, J.R. and Ricard, B. (1995) Generating diagnostic explanations relating to turbine generator failures, in *Proceedings of the 8th International Congress on Condition Monitoring and Diagnostic Engineering Management (COMADEM)*, vol. 1, pp. 385–90.
28. Kirkham, C. and Harris, T.J. (1995) Development of a hybrid neural network/expert system for machine health monitoring, in *Proceedings of the 8th International Congress on Condition Monitoring and Diagnostic Engineering Management (COMADEM)*, vol. 1, pp. 55–60.

7

Alarm processing

David J. Young and James R. McDonald

7.1 INTRODUCTION

Alarm processing can be simply defined as any means used to transform the raw alarm information derived from a power system's control, monitoring and protection functions into a form that is finally presented to the power system operator. This processing need not necessarily imply any use of computer technology to provide 'intelligent' processing of alarm messages. Indeed, alarm processing has been in existence in one form or another for longer than computing technology has been widely available to power systems engineers.

The term 'alarm processing', therefore, encompasses a wide range of techniques which all have one common ultimate aim: to present organized and timely information to the operator concerning the events which are occurring on the power system network. However, as it will become clear, stating the role of alarm processing in this general manner disguises the fact that modern 'intelligent' alarm processing techniques have an additional problem to address, hence their role, and what they ultimately achieve is considerably beyond the scope of that tackled by 'traditional' means.

The problem being alluded to above is that of alarm volume, which is ultimately a result of various trends in modern power systems development.

1. The power system networks themselves are larger and more complex.
2. There is a trend toward control and monitoring being performed by small numbers of large control centres.
3. The improvements and reduction in cost of remote monitoring and control technology, in conjunction with similar advances in communications technology has vastly increased the amount of information gathering which can be carried out at remote substations.

All these factors have contributed to a change in emphasis with respect to the role that alarm processing has to perform. It is a case of the technology of alarm processing having to counteract the effects of advances in other areas of power system technology and development. Therefore, nowadays, alarm processing has an increasingly important role to play in combating the potential for information overload which the power system operator could experience, especially under conditions of power system disturbance. Such conditions, when they occur, can cause a flood of alarm messages to be received at the control centre and operators have to attempt to sift through this deluge of information to determine firstly the state of the power system, and secondly where and when action has to be taken to return the system to normal operation. In extreme circumstances, such as those which occurred during the severe storms in the South of England in October 1987, power system control centres were deluged with alarm messages to such an extent that making sense of them immediately was impossible. Operators simply could not keep up with the influx of messages, so their understanding of the current state of the system was lost and hence their capacity to control the power system. An alarm processor which could have filtered and summarized incoming alarm messages would have considerably helped in pinpointing areas where immediate action should be taken to alleviate the effects of the collapsing system. This would in turn have significantly accelerated the task of returning the power system to full operation.

Clearly then, alarm processing has an increasingly important role to play in modern electrical power system control and could ultimately provide tangible cost benefits through minimization of system unavailability caused by serious fault conditions or power system emergencies.

7.2 THE HISTORY OF ALARM PROCESSING

The degree of processing carried out has improved in sophistication over the years, bringing progressively more useful presentation of alarm information to the operator. The techniques which have been adopted are described in the following sections.

7.2.1 Message routing

In order to provide a naturally distributed workload within a central control room facility, the power system network is subdivided into geographical control areas which are the responsibility of a particular operator. Each operator has a control console, where alarms emanating from his or her allocated control area are directed. This routing of messages may be defined and fixed at the installation of the SCADA system, but more often nowadays, the operator is afforded some manual control over the messages which can be routed to his or her console. This allows sufficient flexibility for an operator to manage any allocated control area from any of the consoles within the control centre.

7.2.2 Message prioritization

Clearly there are wide variations in the importance of alarm messages which are received at the control centre, and it would be appropriate to have a mechanism through which the differing importance of messages can be conveyed to the operator. The simplest means is to define several fixed levels of priority into which all alarm messages are allocated. When a message is received at the control centre its priority level is determined by the SCADA message processor and it is displayed accordingly on the operator's console, with colour, flashing text or even audible signals being used to distinguish between messages of differing priority. Such a scheme operates successfully during normal power system conditions, but clearly, during serious fault conditions the operator is likely to be deluged with many high-priority messages, and will have little interest in attending to those of lower priority. Therefore, most SCADA systems provide manual control over the levels of alarms which are displayed at the operator's console. The form of control offered could take two forms.

1. The ability to manually downgrade all alarms by one level, such that the special audible or visual annunciation of high-priority alarms is disabled: this is particularly useful during emergency conditions where the flashing or audible signals provided by many high-priority alarms would be a considerable distraction to the operator.
2. The ability to filter out alarms below a particular priority level: this will simply prevent display of the less-important alarms while the operator attends to higher-priority alarms only.

7.2.3 Advanced alarm-processing techniques

The approaches discussed above represent the conventional mechanisms used in modern SCADA systems to provide power system operators with alarm messages in an ordered and logical form. Given the capabilities of modern hardware and software, and using technologies such as IKBSs there is considerable scope to improve upon these techniques in two main areas:

1. truly dynamic message prioritization;
2. real-time event analysis and alarm summarization.

In the first of these, the alarm-processing system automatically responds to changing power system conditions, rearranging relative message priorities appropriately. For instance, during a power system emergency when the arrival rate of alarm messages is very high, the system would automatically lower the priority of routine warning or informational messages. As a result, only messages relating to fault events on the power system would be presented to the operator.

Implementation of such a scheme depends on two main factors. The first of these is the identification of credible criteria for recognizing when the power system has entered an emergency state and when it has returned to a normal state, such that the alarm processor can be programmed to act on these conditions and change its prioritizing mode appropriately. Basing such criteria on rate of alarm arrival, or perhaps more accurately, rate of fault-related alarm arrival could be stated as a sensible approach, but it should be noted that establishing exactly what levels constitute the changeover points would require careful thought and much fine tuning. The second factor is the definition of a sensible prioritization scheme for normal operation and its corresponding scheme during system emergency conditions. Again the main problem is not in the implementation itself, but in the establishment of agreed priority levels, and what exactly constitutes an 'important' message.

Although the implementation of dynamic prioritization may be relatively straightforward in itself, the degree of improvement in alarm presentation which it provides to the operator falls short of that which is potentially offered by the second technique: real-time event analysis and summarization.

Here, the alarm processor infers meaning from alarm patterns and prints out a message to the operator stating the actual event on the power system which the alarms might refer to. The main benefit offered over dynamic alarm prioritization is that as well as effectively reducing alarm message volume by summarizing several alarms in one concise message, that message can be made to convey interpretive, higher-level information about real events that are occurring on the power system. For example, an earth fault on a transmission line might typically produce several circuit-breaker and relay trip messages, which would be summarized in a one-line message stating the inferred location of the fault and confirming the correct or incorrect operation of the protection system. With the alarm processor providing such high-level summary messages, the cognitive load imposed on the operator is greatly reduced, and he or she sees immediately a picture of the events taking place on the system rather than having to sift through raw alarm information to piece together these conclusions him/herself.

Clearly now, there is some commonality between the aims of the summarization approach to alarm processing and the field of power system fault diagnosis. Indeed the two fields of development will inevitably converge at some point in the near future (Figure 7.1).

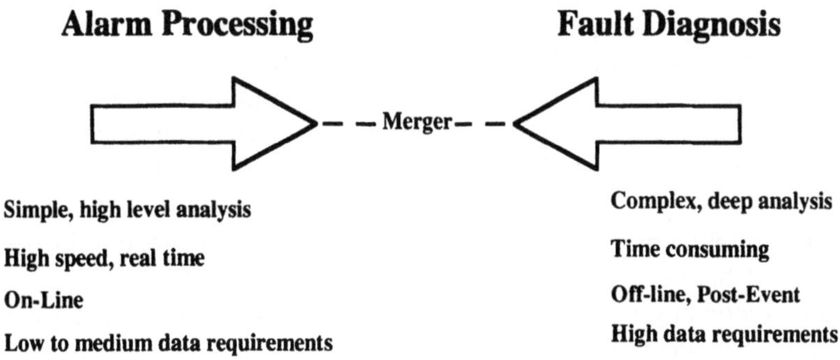

Figure 7.1 Alarm processing and fault diagnosis.

At present, fault diagnosis implies a very complex analysis which uses sources of data such as fault recorder information in addition to alarms and other telemetry in order to determine the location and type of faults, how they occurred and to identify protection maloperations. The event-summarization approach to alarm processing is targeted more at providing on-line, real-time messages to the operator, as and when events occur, and makes simplified assumptions about the correct operation of the protection system. In essence it is fault diagnosis at a shallow level, optimized for performance rather than depth or comprehensiveness. However, with the progress of research in both these areas, in conjunction with increases in the computing power available in control centres, and the depth of on-line alarm data, the alarm processor will tend towards the full capabilities of fault diagnosis. By the same token, fault diagnosis will become feasible as an on-line, real-time application.

7.2.4 Review of previous work

Much of the original motivation among researchers for introducing IKBSs into the alarm-processing domain came from the Electrical Power Research Institute (EPRI) study [1]. This was a human-factors-oriented review of current operations in electrical power dispatch control centres and identified the significant problems presented to operators by the increasing volumes of alarm messages which they could be faced with, particularly during power system emergencies. The EPRI report recommended that: transformed data be presented to operators rather than raw data; the transformation should improve the operator's view of the power system; some effort be directed toward dynamic alarm prioritization. Wollenberg's paper [2] made reference to the EPRI study and set out to describe a feasibility study and initial practical experience of developing a prototype intelligent alarm processor. This system was based on LISP and was aimed at filtering out duplicate and superfluous alarms as well as implementing rudimentary summarization messages. The paper also demonstrated the potential alarm-volume reductions possible through the intelligent alarm-processor approach via a number of worked examples of operation of the prototype system. These showed that the intelligent alarm processor could achieve reductions in alarm volume of between 40% and 80%. Clearly, this served as an important initial demonstration of the true potential of what IKBS-based alarm processing could be expected to achieve.

Following soon after Wollenberg's paper, a number of other papers appeared which emphasized the importance of artificial intelligence (AI) applications in power systems in general, and alarm processing in particular. The first of these by Amelink *et al.* [3] described the existing alarm-processing techniques and their shortcomings and called for the use of AI to provide the great improvements required in alarm processing to help alleviate the load on operators. Particular mention is made in this paper of the use of AI techniques to provide dynamically adaptive prioritization of alarm messages such that the relative priority of messages depends on current power system conditions. Wollenberg contributes to two further papers advocating the use of AI in power systems. The first in collaboration with Sakaguchi [4] shows the differences in development methodology between AI program development and conventional algorithmic program development. It then goes on to address the alternative means for incorporating AI processing into the existing SCADA system computer facilities. Three possible approaches are given:

1. Implement AI software in an AI language such as LISP and run it on the energy management system (EMS) computer.
2. Implement the AI program in a conventional language such as Pascal or C and run it on the EMS computer.
3. Implement the AI program in LISP and run it on hardware dedicated to running LISP code.

Wollenberg and Sakaguchi describe the advantages and disadvantages of each approach but, quite wisely, do not commit themselves as to a preference.

The second collaborative paper in which Wollenberg was involved [5] first offers a comprehensive list of potential areas of IKBS application in power systems. There is then a description by Larsen of the early proposed programme of development for the NSP (Northern States Power) Intelligent Alarm Processor. Wollenberg concludes the paper by emphasizing three important prerequisites for AI being useful in an EMS environment:

1. high-speed execution of the software, so that real-time applications such as alarm processing are feasible;
2. the ability for the AI software to be coupled easily to the existing EMS facilities, including databases and human–machine display and control facilities;
3. support of the AI software by facilities for ease of development and maintenance to provide at least the level attained on stand-alone AI development workstation systems.

The 1989 paper by Prince *et al.* [6] was important in that it brought to light the true needs of the community of power system control centre operators, and coined the phrase 'one event, one alarm'. This again emphasized the need for intelligent alarm-processing techniques which could reduce the many alarms produced during a single power system event such as an earth fault on a line down to a single event summary message, thus greatly reducing the message volume presented to operators, while at the same time satisfying the aim of presenting them with interpretive higher-level information.

A number of papers appeared from 1989 onwards describing on-going alarm-processor developments at various electric-power utilities in the USA. One of the more interesting papers is by Tesch *et al.* [7] who describe the system they have developed in association with Wisconsin Electric Power Company. Their alarm processor is based on the C programming language, which is justified by claims of high performance, and the use of a rule-preprocessing approach where the rules which drive the system can be developed and added to without altering the C program code of the alarm processor itself. The system also uses a novel combination of a forward and backward chaining strategy in order to derive conclusions about the events occurring on the power system.

Another interesting system is the NSP Intelligent Alarm Processor described in the paper by Bijoch *et al.* [8]. The paper was written after several months of active experience of operating the alarm processor in Northern States Power's control centre. Like the Wisconsin system, this alarm processor has been a custom development and has not used LISP or any of the commercially available IKBS tools or shells. Two further important points of some note are raised in the paper. First of all, and very significantly, this alarm processor has been permitted to perform automatic control actions according to the results of the analyses it carries out. At present this is limited to capacitor switching depending on voltage conditions on the network, but the implication is that further control actions may be added in future. Secondly the paper gives some important information on the degree and extent of testing carried out on the system to ensure its integrity and suitability for real on-line use. Shoop *et al.* [9,10] describe the SOCCS (System Operation Computer

Control System) Alarm Advisor (SAA) which was put into operation by Consolidated Edison of New York in June 1989. The SAA is an on-line real-time IKBS which performs nuisance alarm suppression and provides recommendations to operations staff on actions to take based on its analysis of incoming alarms. Model-based reasoning is employed based on functional, temporal and topology models.

Model-based reasoning and heuristic techniques are combined in the system described by Pfau-Wagenbauer and Nejdl [11] which is integrated into a large SCADA system for power system distribution networks. The IKBS itself is based on KEE (Knowledge Engineering Environment) and runs on a workstation which is linked into the main SCADA computer system via a local area network. It provides operator support by diagnosing disturbances and device malfunctions on the 22 kV distribution network run by the Public Utilities Board of Singapore, and has been in use since November 1990.

The paper by Kiernan and Warwick [12] is particularly interesting in its use of very novel 'genetic algorithm' techniques for adaptive on-line diagnosis of power transmission network faults. The system monitors switchgear indications and reports fault diagnoses on discovery of any patterns indicative of faulted components. The system evaluates the accuracy of diagnosis via a fault simulator developed by National Grid Co., and adapts to reflect the current network topology through use of the genetic algorithms approach. The principle behind this novel concept is that the success of a particular algorithm is used to perform an evolutionary natural selection process akin to that which occurs in nature, and hence over time the system becomes better at performing its job. The use of genetic algorithms for alarm processing is also discussed by Wen and Chang [13].

There are other papers worthy of note, which are not directly associated with power system alarm processing.

Cardozo and Talukdar [14,15] present their fault diagnosis IKBS which is of particular interest due to its novel 'distributed problem solving' (DPS) methodology. By implementing a network-based distributed architecture for their alarm processor they have provided one possible solution to the significant processing demands posed by deep diagnosis of alarm messages.

Hemmelman *et al.* [16] describe a communications alarm processor in use at Bonneville Power Administration. This alarm processor is being used to handle alarms relating to the operation of the company's extensive data acquisition, voice communications and SCADA network, rather than the operation of the power system itself, but clearly the practical problems and techniques being used to combat them are the same as for power system alarm processing *per se*. It is interesting that in this paper a number of pertinent points are raised concerning the cost benefits which the alarm processor is expected to provide particularly due to savings in maintenance-related activities. The same expected benefits could be said to apply to power system alarm processing, yet there is little mention of this in the other literature.

The nuclear industry is another area in which alarm processing has strong potential for application. Many of the same principles apply here as they do towards power system alarm processing, and the two papers by Cheon *et al.* [17,18] provide interesting points for comparison. The first paper is particularly interesting in that the feasibility of an alarm processor based on an artificial neural network is discussed. The second paper describes a relatively more conventional alarm-processing IKBS called ESAPD (expert system for alarm processing and diagnosis), the main purpose of which is to assist operators in identifying the primary causal alarm amongst multiple-fired alarms, and hence to diagnose more quickly and directly the plant malfunction being indicated.

For general considerations relating to the form and presentation of alarm messages the papers by Corsberg [19] and Danchak [20] are worthy of study.

7.3 THE PRACTICAL PROBLEMS OF ALARM PROCESSING

There are a number of practical problems inherent in the alarm-processing task which have to be overcome when developing an alarm-processing IKBS. These are:

1. achieving adequate speed of operation;
2. the limited scope of telemetry;
3. missing alarm messages;
4. unreliable time ordering of alarms;
5. the unpredictable arrival rate of alarms;
6. simultaneous occurrence of many separate events.

The specific nature of these problems is discussed in the following sections.

7.3.1 Speed of operation

Alarm processing is a demanding real-time task which requires alarm-processing software to keep up with very high rates of alarm arrival. Typically, for a large SCADA system, the potential exists for 500 alarms per minute to be received at the control centre. In effect this means that at such a peak arrival rate, the alarm processor must take no more than 0.12 s to process one alarm, perform relevant analysis and be ready to receive the next alarm message. Therefore in constructing a practical alarm-processing expert system, speed of operation becomes a primary concern, and influences many aspects of the design of the resulting system.

7.3.2 Limited scope of telemetry

In order to perform intelligent processing of alarm messages, the IKBS relies very heavily on the information inherent in the telemetry messages it receives from the power system's SCADA network. In an ideal situation, every item of plant in the power system network would transmit appropriate data such that the alarm processor would be able to ascertain its current state for the purposes of performing a correct analysis of the system. Items of particular significance would be circuit breakers and protections, as detailed live information on the states of these would allow the alarm processor to provide very accurate analyses of fault situations. However, it is not economically viable to transmit every single point in a power system network. Indeed, often the lower voltage levels of a distribution network carry no telemetry at all, and even at the intermediate levels such as the 33 kV level, telemetry of circuit breakers and protections is by no means comprehensive. It is only at higher, transmission level, voltages that quality of telemetry approaches the ideal, but at the same time, the complexity of the network above distribution level diminishes and correspondingly the need for alarm processing to aid the operator becomes less significant. Therefore, since distribution networks have the greatest need for alarm processing, but tend to have incomplete telemetry, it is a fact that most potential alarm-processor applications will have to be able to perform their analysis in the face of only a partial picture of the true power system state.

7.3.3 Missing alarm messages

In addition to the lack of completeness in installed telemetry which the alarm processor must cope with, an additional problem arises through the relative unreliability of alarm messages. There are many reasons why alarm messages may not reach the control centre: a faulty contactor in an alarm circuit, a faulty substation data acquisition computer, communications link problems etc. Any of these may prevent a message concerning the action of some item of substation plant or protection being communicated back to the control centre. The alarm processor must therefore be aware that some of the evidence it is seeking in order to derive a conclusion may not in fact arrive, and therefore it must be able to take this into account in weighing up possible solutions. This of course adds a further dimension of complexity to the analysis task the alarm processor must perform – and indeed is one significant area to which the 'intelligent' processing offered by an IKBS approach is well suited.

7.3.4 Unreliable time ordering of alarms

A further problem which the alarm processor must confront relates to the ordering of alarm messages. In general the order of messages within the alarm stream does not always reflect the actual order of events which have taken place. This is due to the way in which the information is gathered within the substation and the way it is transmitted back to the control centre. If, for example, protection at the substation sees an earth fault on a feeder, it will cause the appropriate circuit breaker(s) to be tripped in order to isolate the fault. As a result the station data acquisition equipment may typically have the protection operation and the breaker open indications flagged in its memory. Note that even at this stage the order in which these flags were raised will not be recorded unless the system is capable of 'time stamping' (the attaching of time information to the actual event information) of events at this level. Some time later (as much as 5 min depending on the polling interval) the control centre SCADA computer will poll the substation computer to request any new state changes which have occurred. At this point the protection status and breaker status messages are transmitted back to the control centre. The order in which this is done depends more on the software/hardware set-up of the data acquisition equipment and the central SCADA computer than the actual order in which the events occurred. Another consequence of the polling scheme is that the order of polling of substations governs the order in which events occurring virtually simultaneously at different substations are annunciated via alarm messages on the operator's console.

Therefore, unless time stamping of messages at source is a facility of the SCADA system, the alarm processor cannot make any reliable assumptions during its analysis based on the order in which it sees alarm messages arriving. In other words, the alarm processor's analysis must be able to be carried out and produce the same conclusions independent of the time ordering of the alarm messages.

7.3.5 Arrival rate of alarms is not constant

As a further consequence of the way in which remote data is gathered, messages relating to an event may arrive with significant time delays in between. For example, if an event commences just prior to the control centre polling the substation computer, then any alarm messages caused by the commencement of the event will be transmitted back immediately, but a full polling period will have to elapse before any further messages relating to the event are gathered. Hence it is quite possible for a protection operation alarm to be received and a full 5 min elapse before the expected breaker trip indications are seen at the control centre.

The obvious problem which must be overcome by the alarm processor is that it must be able to perform an analysis spread over several polling periods, and not rely on all the data relevant to a particular event arriving in one convenient block.

7.3.6 Multiple simultaneous events

Clearly, it is quite possible for many events to be occurring on the power system network simultaneously. This would be especially true during severe weather conditions where the power system may be subject to many fault events within a relatively short period of time. It is therefore essential that the alarm processor be able to perform analysis of several events at the same time, given that the analysis of one event alone may often require the examination of alarms occurring over several polling periods. If the alarm processor were not able to progress the analysis of several events at the same time there would be very significant delays before it produced any messages on the operator's console.

In effect the alarm processor should not adopt a purely serial approach to its analysis of events, but instead should adopt a parallel approach, where many events can be considered and progressed at the same time, and conclusions derived as and when the appropriate evidence arrives. This means that at any time during operation the alarm processor will be holding several analyses in progress, with new ones being formed, and old ones being completed as time goes on.

The foregoing discussion of these practical problems highlights the areas which must be tackled satisfactorily in order to produce a practical IKBS for alarm processing. These problems are peculiarities of the alarm-processing application which make it different from many other applications of IKBSs. Indeed, many of these practical problems render the simple, well-known rule-based IKBS techniques and tools unusable in an alarm-processing role, and more novel approaches must be sought in order to produce a viable solution.

In the next section the actual process of building such a system is considered, with all the above problems being borne in mind.

7.4 DEVELOPMENT OF AN ALARM-PROCESSING EXPERT SYSTEM

The real-time alarm analysis and summarization approach to alarm processing is well suited to implementation using IKBS techniques. The overall methodology of implementation is shown in Figure 7.2.

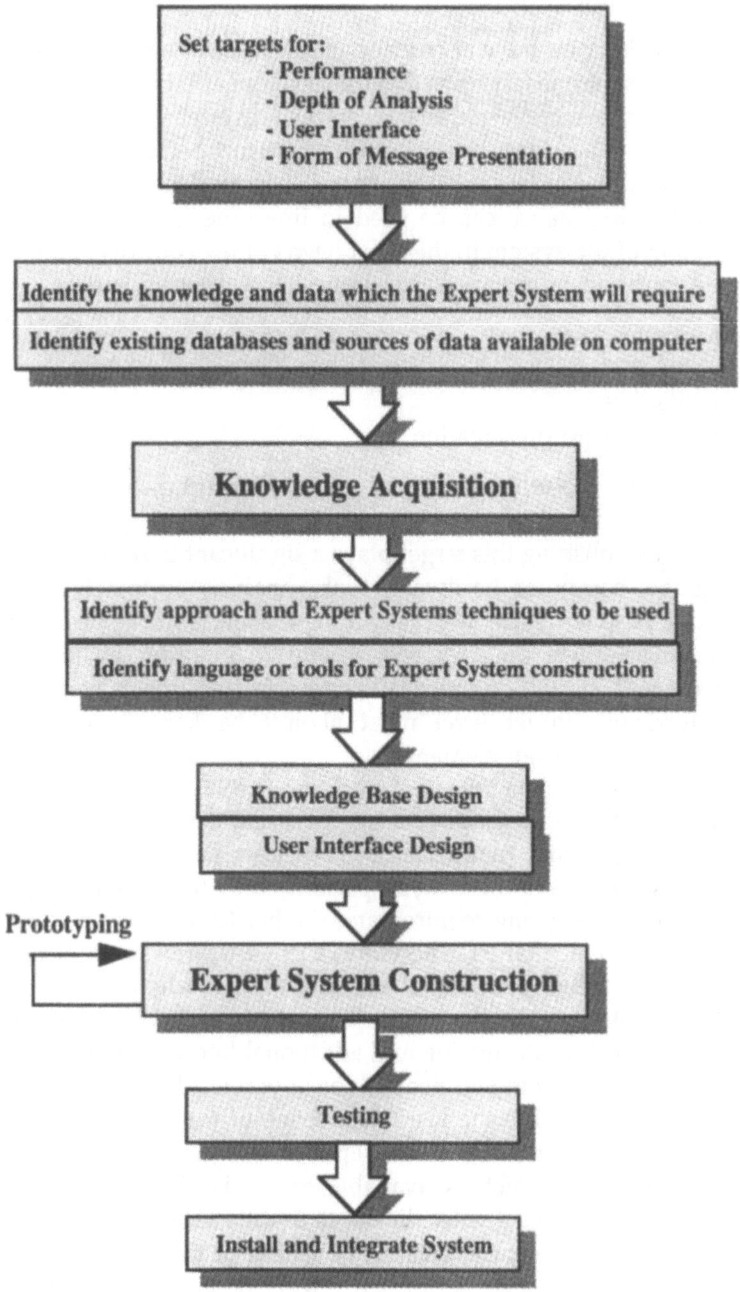

Figure 7.2 Overall methodology.

7.4.1 Setting initial targets

Before embarking on the construction of an IKBS for alarm processing it is essential to set some targets so that the developer of the system has some definable aims to achieve within the completed system. These targets will necessarily take account of user (power system operator) requirements but these need not at this stage be regarded as final and definitive, since later prototyping stages can be used to fine tune many aspects of the presentation of the system to the exact needs of the control engineer.

Four significant target areas are suggested for consideration:

1. performance
2. depth of analysis
3. user interface
4. form of message presentation.

The first target, performance, is of primary importance. It is necessary to quantify the levels of alarm arrival which the alarm processor must cope with. Establishing this target plays a significant part in the selection of language, computer hardware, tools, analysis approach and even knowledge base structure, because all these aspects of the design of the IKBS have an effect on its speed. Therefore in order to ensure that the resultant IKBS will be practically operable in terms of performance, the performance target level must always be kept in mind when addressing these design elements.

Similarly, the depth of analysis which the IKBS is expected to perform plays a very large part in governing the overall design of the system. Depth of analysis and performance go hand-in-hand to a large extent, since clearly the deeper the analysis performed by the system, the greater the amount of processing required and the harder it becomes to achieve a given performance target. Therefore, a great deal of thought must be given to the depth of analysis which can be practically achieved within the time constraints on the IKBS. The depth of analysis target also governs how much information and additional knowledge the IKBS will have to have at its command. Unless considerable additional investment in telemetry falls within the justifiable scope of the IKBS development, the system must achieve its given depth of analysis within the bounds of the information which is available to it via the existing SCADA system. In addition, the greater the depth of the analysis to be performed by the alarm processor, the greater the extent of the knowledge acquisition task which must be performed during IKBS construction.

Nowadays the way in which a system presents its information to the operator and how it interacts with the operator is recognized as being of extreme importance. A system may be very capable indeed in terms of advanced and complex in-built facilities and able to display extremely useful information to an operator, but if its user interface is not well designed, the system may be all but operationally useless. It is therefore very important that some targets are set on the system's 'usability', and human factors aspects of the system are taken fully into account from the outset. A successful IKBS is one which is easy to use, and is fully accepted by the operators. By setting targets for what the user interface must achieve, a number of design decisions are affected, such as choice of hardware, operating system, expert system tools, window system, and degree and form of operator interaction expected during analysis. Note that use of a window-based environment of some sort is not a prerequisite of a good user interface. Indeed, many very bad user interfaces arise through poorly considered use of window-based systems. However, it is accepted that in modern user interfaces, well-designed window-based approaches can achieve an efficient, flexible and convenient operating environment for the power system control engineer.

The fourth important initial target to define is the form of message presentation to be used. If the aim is to provide a single-line summary message then other points must also be considered: should this message literally always be a maximum of one line, or will more detailed explanatory messages be required? It may in fact be a requirement of the system that it has the facility to provide some indication of its reasoning in reaching a particular conclusion. Such a facility is very valuable both from the point of view of the information it provides, and the way in which it helps to increase operator confidence in the system. Another point to consider is whether the system should present only one possible solution or in the case of ambiguity in input data, a selection of the most likely possibilities.

7.4.2 Identification of knowledge and data requirements for the IKBS

Having set the initial targets for the IKBS, the next stage in the development process is to identify the knowledge and data which it will require to perform its defined role. The depth of analysis which has to be performed clearly affects the scope of the knowledge and data which the system will require to work upon.

The most obvious data requirement which the alarm-processing IKBS will have is access to the real alarm message stream. During development and testing of the system it is not often practical to have access to a live SCADA system message stream, but instead very effective test situations can be built up using data files containing recorded power system alarm streams.

The next major data requirement is access to a database which describes the power system network topology. This contains data describing the connectivity between major items of power system plant, such as lines, transformers, busbars and circuit breakers. More detailed topology databases may also contain information on the location and type of protection equipment. Most power utilities have such databases in place, either as part of a graphical 'geographic information system', or to provide the topology model used by power system analysis packages performing such tasks as load-flow and security assessment.

A third, very important data source which the alarm processor will require is a database describing the alarm and indication messages which are available at each substation in the network. In effect, this data allows the alarm processor to determine the actual alarms it can expect to emanate from a particular substation during a power system event. Without such information the expected alarm sequences for an event could not take account of the actual type of alarms available at the substations.

It should be further noted that if the references to particular substations or items of plant which the alarm processor 'sees' in the alarm stream are not directly attributable to entities within the topology database then the alarm processor requires an additional database to provide that cross-reference information. A simple illustration of this problem would be the case where perhaps an alarm message contains an abbreviated name for a substation, 'NRTHTN' and all references in the topology database use the full name 'NORTHTOWN'. Additional data would be required in order to allow the alarm processor to relate 'NRTHTN' or similar abbreviations to the actual substation name used in the topology database.

Depending on the depth of analysis of power system events which the alarm processor is to perform, other sources of data may be required: For example, data on protection attributes such as type, directionality and reach is essential to make sense of many complex fault situations. Statistical data on power system plant and fault types can be used to refine the IKBS's concepts of what constitutes a 'most likely' event scenario.

So far the tangible, computer-based data sources which the IKBS will require for operation have been addressed, but clearly in addition to these, to perform its nominal role as an 'expert' in alarm processing, the system will require to embody in some way the knowledge about the inter-relationship of alarms and events which a real human expert possesses. The actual structure of this knowledge and how it is operated upon in order to perform analysis of the incoming alarm stream is addressed in later sections.

Many of the points raised in relation to data requirements may be seemingly insignificant in themselves, but they do need to be considered at the outset. As is the case for a human, if the structure and content of the information being fed into the IKBS is not considered carefully in advance, its task in analysing it and coming to useful conclusions is much hindered.

7.4.3 Identification of existing data and knowledge

The previous section addressed the data requirements for the alarm-processing IKBS without concern at that stage for whether those data sources were in existence or not. The next stage of the development process is to identify which of the data requirements can be immediately satisfied by existing databases, and which require further data to be acquired and made available in computer-based form.

Where required data is not already in a computer-based form, it must first be decided if it is indeed feasible to acquire this data and construct the necessary databases. The cost and time required to provide such data may prove prohibitive and the IKBS's continued development may have to be reconsidered. It is clearly necessary to identify, and if possible satisfactorily address, these problem areas before further effort is expended in implementing the IKBS.

7.4.4 Knowledge acquisition

This stage refers to the acquisition of 'expert' knowledge which will guide the analysis performed by the system.

There are several different types of knowledge involved in the building of an IKBS for power system alarm processing. These could be summarized as being of the following types:

1. qualitative general power system knowledge;
2. quantitative general power system knowledge;
3. specific power system knowledge;
4. qualitative general SCADA system knowledge;
5. specific SCADA system knowledge.

Qualitative general power system knowledge is a knowledge of the basic function and interaction of elements of power system networks. For example, it is an appreciation of what circuit breakers do, types and purpose of protection, what transformers do and so on. This is in contrast to the quantitative general power system knowledge which is concerned with the 'mathematical' theory behind the operation of elements of the power system. For example, knowledge of protection characteristics, use of fault calculations etc. An individual possessing a good qualitative general knowledge of power system operation may not possess any expertise at all in the quantitative field, but conversely, the ability to apply quantitative knowledge demands that the individual already possess the qualitative knowledge.

Specific power system knowledge is knowledge of the operation and history of a particular power system network, including any special features it may possess over and above those normally found. Clearly the type of person who has this knowledge is likely to be an employee of the power utility which operates the network in question, and who has had a number of years experience working directly with the development or operation of the network.

Qualitative general SCADA system knowledge is a general appreciation of the principles of SCADA system operation; how data is collected, communicated and then presented at the control centre. This knowledge is general in the sense that it embodies principles that are applicable to all SCADA systems.

Specific SCADA system knowledge is a knowledge of the particular SCADA system involved in the application for which the IKBS is being developed. Again this form of knowledge is only likely to be possessed by employees of the power utility who have direct experience of the development and operation of the SCADA system over several years.

Clearly, the team involved in the development of the IKBS will have to be able to draw upon all these types of knowledge to develop an IKBS for a particular alarm-processing application. Some of the knowledge types could be expected to be possessed directly by the developers themselves, but unless they are also employees of the power utility they are less likely to have any initial knowledge of the specifics of the power network or SCADA system which the alarm processor will be required to work with. There are a number of ways in which the IKBS developers can gather the knowledge which they require:

1. structured interviews
2. documentation
3. questionnaires
4. observation.

Interviewing provides a direct method by which the IKBS developers can extract knowledge from identified experts through a formal question-and-answer session. In order for interviewing to be an effective method of knowledge gathering, the IKBS developers must firstly have enough of an appreciation of the subject domain in order to know the relevant questions to ask of the expert, otherwise much time and effort can be wasted pursuing areas of knowledge which have little or no relevance to the needs of the IKBS. The expert is naturally liable to dwell on his or her own areas of particular interest and is unlikely to convey useful knowledge unless firmly directed via appropriate questioning.

A very useful way of gathering a great deal of detail specific to the operation of a particular power system network and SCADA system is through examination of drawings and documentation. It is important, however, that material is selected carefully for relevance in terms of degree of detail required for the actual IKBS application in question. Documentation therefore is particularly useful in tackling the problem of gathering power system specific knowledge.

Questionnaires can be used to gather information in a well-structured manner from a group of experts. In the case of an alarm-processor development, a questionnaire might be particularly useful in gaining operators' opinions on how the power system and SCADA system will react to given situations, and how they would interpret this information. Similarly questionnaires would be ideal for gathering operators' opinions on how the alarm processor should present its information, and the form its user interface should take.

Observation of operations carried out in the control centre, or by staff at a substation can be extremely valuable in gaining knowledge on the types of situations which may arise, how to recognize them from telemetry and how they are tackled. Observation is also invaluable in clarifying the developers' perceptions of how the power system operates, and how the personnel who manage it carry out their jobs. Such clarification is often required in order to put into true perspective the information which has been gathered by other means or by reading documentation. Only with such first-hand experience can many of the important features of the task be fully appreciated.

Knowledge acquisition is therefore a task of quite considerable scope, which the developers must devote time to, both in an active and a more passive manner during the construction of the IKBS. This entails embarking on specific exercises using approaches, as discussed above, with the particular aim of gathering chunks of knowledge for later incorporation into the system. In addition, the more passive type of knowledge acquisition occurs right through the development process as the developers progressively gain more of a feel for the pertinent issues in the problem area through discussion, and other informal contact with the experts.

7.4.5 Identification of approach and IKBS techniques

Once knowledge acquisition is in progress, the next stage in the development is to examine the problem area in some detail in light of the information gathered up to this point, and determine the best IKBS approach which can be used. There are numerous philosophies of approach which can be taken: rule-based, frame-based (object-oriented), hypothesize-simulate-and-test and even neural networks present some possibilities.

The selection of approach should be carried out while taking into account the data and knowledge resources available, the targets which have been set for the system and, very importantly, the practical problems inherent in the alarm-processing task, as discussed in section 7.3.

From experience, the most fruitful avenue to pursue is the hypothesize-simulate-and-test approach as this is ideally suited to the needs of alarm processing. It has a number of features which enable IKBSs so constructed, to be made to cope relatively easily with the practical problems of alarm processing. In addition, the approach allows for convenient expansion of the capabilities of the IKBS from the point of view of depth of processing performed. Thus, the IKBS can be developed initially to perform relatively simple processing, but as quality of telemetry and speed of processors improves in the future this can be augmented to provide full fault diagnosis capabilities on-line.

7.4.6 Identification of language or tools for IKBS construction

Speed of operation has a significant part to play in the selection of language or IKBS tool to be used for the actual construction of the alarm processor. It is unfortunate that at present many of the available IKBS tools, or languages conventionally used for IKBS construction (e.g. LISP and Prolog) are relatively inefficient in their use of hardware computing resources, and thus do not provide the ideal environment for achieving the speed of operation required for alarm processing. This situation will increasingly be alleviated by improvements in computer hardware, allowing more practical use of LISP, Prolog and IKBS tools based on these languages in real-time applications.

However, at present one possible approach to achieving the high speed required of an alarm processor is to use a conventional high-level language such as 'C' or Pascal. These offer a further significant advantage: they are easily integrated into real SCADA system environments, where the capability to interface with database software and window management software is extremely important. The latter feature is important in that it allows for construction of a suitable user interface. The main disadvantage of using a conventional high-level language is that it does not come with all the commonly used knowledge representation and manipulation facilities which are built into most IKBS tools, and which are naturally achieved within 'symbolic' programming languages such as LISP and Prolog. However, with appropriate programming knowledge in C or Pascal these problems can be overcome, and as long as some care is taken in the way knowledge and data are represented, it can be accomplished in as flexible a manner as LISP, Prolog and IKBS tools offer.

7.4.7 Knowledge base design

There are two important aspects to the design of the knowledge base:

1. how data and knowledge are represented and handled internally;
2. how data and knowledge are presented externally.

To clarify the distinction between these: the internal representation is the form in which the knowledge is held inside the IKBS, and the form in which the IKBS manipulates the knowledge when using it to perform an analysis. The external representation of the knowledge is the form in which it is entered, viewed and maintained by the developer. Clearly if one form is used in both situations there must be a compromise between usability of the information by the human on the one hand, and on the other hand, the performance that can be expected in manipulating the data internally by the computer. Data that is ideally suited to a human in terms of readability and understandability is normally very bad from the point of view of efficiency of storage and manipulation by the computer and *vice versa*.

One of the features of symbolic programming languages such as LISP and Prolog, and IKBS tools, is that the internal and external representations of the knowledge are very closely related, if not exactly the same. For example, LISP's natural data representation structure is a 'linked-list', and a rule or a LISP program itself will be held internally as a linked-list. In addition, the rule or program is actually structured like a list in its textual form when inspected visually. Similarly, in Prolog the natural way to structure everything is as a statement in predicate logic. Although both these approaches are powerful from the point of view that the language is constructed around one consistent form of data structure, there are significant disadvantages. Firstly, not all forms of knowledge or data which the IKBS might potentially need to manipulate are naturally well suited to representation in the structures imposed by these languages (there may be significant performance penalties imposed by forcing data into a linked-list layout when it might be better held in a more straightforward structure such as an array for example), and secondly because the external representation of the data is in fact laid out in the manner appropriate to the programming language in question it is not ideally suited to viewing, updating or checking by anyone other than an individual reasonably familiar with LISP or Prolog.

A good approach to adopt, especially if using a high-level language for implementation of the IKBS, is to design a suitable external structure for data and knowledge which is in a convenient, logical, human-readable form. This can be optimized to be as easy to use by a human as possible, and at the same time, by translating it into a convenient internal form automatically, the same data can also be made optimally usable by the IKBS. Clearly the implementation of the program code to perform this automatic translation between the external and internal forms of the data involves more programming effort but the techniques involved are relatively straightforward. The results of taking this approach provide significant performance advantages while not compromising the usability and maintainability of the knowledge base for the human user.

7.4.8 User interface design

A very important part of the IKBS is its user interface. Some time and care should be taken over the design of this, since it will be a major factor in the acceptance of the system by the power system control centre operators. In order that the user interface is designed to suit their needs operators should be consulted and involved in the process of development of the user interface. Many aspects have to be considered such as sizes and colour of text, use of windows, the order in which messages are displayed by the user interface, and the design of any controls which the operator will require in order to interact with the IKBS.

7.4.9 IKBS construction – knowledge base and user interface

Once all the previous stages have been completed, the actual construction of the alarm-processing expert system can be carried out. A phased approach to the development of the system is an ideal way of achieving early results which can be built upon, and refined progressively towards the final system. This of course ties in well with a prototyping approach, where the completion of each phase of development corresponds to a milestone where the system is presented to the end user for comment. A suggested breakdown of the system development might fall into the following phases:

1. coding of the alarm message stream input and analysis module;
2. direct coding of knowledge structure in the implementation language;
3. development of separate external rules/knowledge structures;
4. initial user interface;
5. expansion of external rule base.

Stage 1 produces the rudimentary basis of the system, which is able to read and break down a stream of alarm messages into its constituent parts, such as time, location of the alarm, type of alarm and alarm state. These are the fundamental pieces of information about each alarm which the alarm processor will then be developed to understand and act upon. Stage 2 would take the system to the point where some simple rules for the recognition of a small subset of power system events are coded directly into the alarm processor itself. This allows very rapid production of a system which can be used to demonstrate reaction to a limited set of scenarios. The alarm processor's mechanism for recognizing events is therefore in place but is implemented in an inflexible manner. The next stage therefore is to begin the process of developing the final external representation of rules such that they can be maintained as entities separate from the alarm-processor code itself, thus enhancing the flexibility of the system immensely. Following this stage, effort could be directed toward development of the user interface, thus taking the system visually closer to its final form and offering another opportunity for demonstration and feedback from the end user. The last stage of development would entail an expansion of the rule base to cover a much larger set of possible scenarios, and if necessary a refinement of the system's analysis mechanisms to cope with more complex practical situations.

7.4.10 Prototyping

Demonstration of prototypes of the IKBS to operations staff, and the resultant feedback gained through this, can be used to iterate the knowledge base and user interface design and construction phases towards their final form. Making use of prototyping ensures that when the IKBS is finally installed and made available for use by the power system operators, it is suited to their needs and their acceptance of it is likely to be guaranteed.

7.4.11 Testing

During development of the IKBS, it is very important to be able to perform realistic tests on the validity of the analysis the system is performing. In the alarm-processing domain, the obvious approach is to use streams of recorded real alarm data as test cases for processing by the system. This is indeed a very valuable method of showing the system's reaction to real alarm streams, with all the inherent peculiarities of that form of data, and goes a long way toward proving how the alarm processor will react when linked up to truly live alarm data from a real SCADA system. It is, however, important that the alarm processor is not built purely around analysis of the particular cases held within the available recorded alarm streams. These will by no means cover all possible scenarios, so the developer of the IKBS must use recorded alarm streams sensibly, devising other test scenarios to prove the generality of the analysis being performed. For example, if a recorded test-case scenario provides the alarms occurring during a particular type of fault, and the alarm processor is able to diagnose the situation correctly from

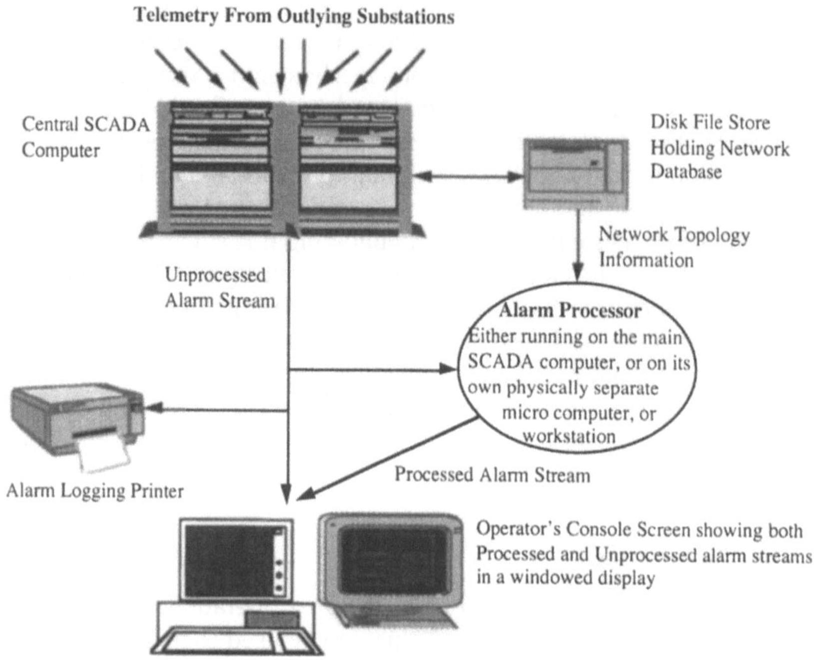

Figure 7.3 Typical implementation of the alarm processor.

these, the tester should go further by observing the system's response when one or more of the alarms is removed from the stream, or the order of the alarms is altered and so on. These simple measures help to prove the IKBS's robustness in the face of the sort of variations in actual messages which might be observed in reality, and prove the independence of the solution from effects such as order of alarms and missing alarms.

7.4.12 Installation and integration

Once the alarm processor has been proven adequately through testing it is ready to be installed into the real SCADA system environment. This will involve work to put in place mechanisms for feeding the live alarm stream into the input stage of the alarm processor, as well as, for example, providing access to actual system topology databases held on the SCADA computer. Figure 7.3 shows how the alarm processor might typically be integrated into the existing SCADA system.

Other integration work will be involved in providing access to the alarm processor's message display and operator control and input functions from the SCADA system display consoles in the control centre. This latter task can be considerably eased if the environment being used is based upon a standard window system such as X-windows, where the incorporation of windows from diverse applications on to the one display is very easily accomplished without the necessity for any interdependence between the applications. In other words a number of self-contained applications can co-exist on the same display without any of them having to be specifically programmed to do so.

7.5 A CASE STUDY: APEX

This section presents a description of a particular alarm processor development called APEX (Alarm Processing EXpert System), and shows how the considerations detailed in the previous sections impacted on the design and construction of that system. APEX is currently targeted for use in distribution control centres, primarily concerned with 33 kV distribution networks but its flexibility makes it easily adaptable to application at other levels.

APEX has the following features.

1. It is implemented in the C programming language. This was chosen in preference to alternative development environments such as Prolog, LISP or a commercially available IKBS shell for the reasons of speed, portability and flexibility discussed previously.
2. The alarm processor adopts a hypothesizing approach as its mechanism for analysing alarms and coming to conclusions about their meaning.
3. It was judged that presenting summarized messages to the operator was the most appropriate means of conveying the same information contained in raw alarms, yet in a much more concise and comprehensible form.
4. Particular effort was directed towards separating the rules which the alarm processor acts on from the program code of the alarm processor itself.
5. Emphasis is given to the overall user interface which the system presents to the operator. Clearly the ultimate value of the system to the operator depends greatly on how convenient it is to use. Observations on the presentation of alarm data have been drawn from relevant work by other researchers [6,18,19], as well as from the construction of a brief questionnaire used to gauge operators' individual preferences.

7.5.1 Details of APEX construction

(a) Knowledge base structure

Early development of APEX involved the direct coding of knowledge concerning the recognition of power system events into the C program source of the alarm processor itself. However, it was very clear that this was a situation which could only be tolerated in the short term since the flexibility and maintainability of the alarm processor would be seriously compromised by such an arrangement. Therefore, effort was directed toward the development of an external form of representation of the knowledge. These could loosely be termed as 'rules' for the recognition of events from characteristic alarm patterns, and they are expressed in a form which is designed to be concise, yet easy to understand even for someone with little programming background. An example APEX rule follows with an explanation of its meaning:

```
Event "Successful reclosure after fault on <BlackOut>"
Expect
{
Alarm AUTOTRIP OPERATED <StationSet>
Indication RECLOSED <CBSet>
}
```

This means that for the event in which a successful reclosure occurs after a transient fault on some part of the network, the expected pattern of alarm messages would consist of an `AUTOTRIP` alarm from each of the substations at which protection operated and a series of breaker reclosed indications from the appropriate breakers. `CBSet` is understood by the alarm processor to mean a set of one or more circuit breakers which would be required to isolate the area `BlackOut`. A `StationSet` is a set of one or more substation names corresponding to the stations where automatic protection action has tripped circuit breakers in `CBSet`. The alarm processor uses the rule by expanding the terms `CBSet`, `StationSet` and `BlackOut` into actual names of circuit breakers, substations and network nodes according to the context of the alarm messages it is analysing. This process is illustrated later in a worked example of APEX operation.

In addition to the terms `CBSet`, `BlackOut` and `StationSet`, APEX also recognizes `CB`, `Line`, `TF` (for transformer), `Station` and `BB` (for busbar). A large selection of known alarm and indication states is also known to the system: e.g. `CLOSED`, `OPEN`, `OPERATED`, `RESET`, `OUT_OF_SERVICE` and so on. With such a vocabulary to describe alarm patterns, it is possible to rapidly add rules for new types of events simply by specifying a characteristic expected set of alarm messages for that event in the format shown in the example.

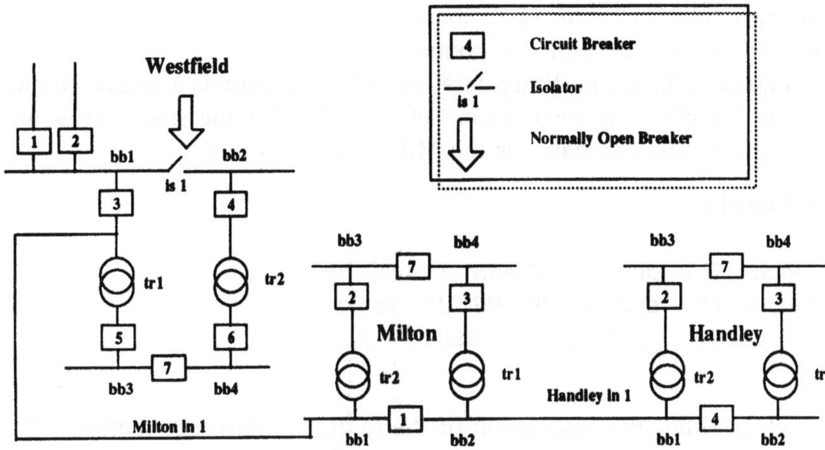

Figure 7.4 Example network.

(b) Structure of network topology database

Like the database of rules, the network topology database must be structured in a manner which is easy to read and maintain. In the case of APEX, the system was developed in the absence of an existing topology database structure, so an appropriate structure had to be designed. A suitable notation for this database was thus developed and it provides a simple means of specifying the topology data in a clear human-readable form. For example, the entry for the Milton substation shown in Figure 7.4 is:

```
MILTON ; Abbreviated substation name
MILTON ; Full substation name
{
; Connection of two halves: busbar 1 and 2

bb1, cb1, bb2.
; Connection of two halves: busbar 3 and 4
bb3, cb4, bb4.

bb1, tr2, cb2, bb3.
bb2, tr1, cb3, bb4.
bb1, ln1.
}
```

The physical connectivity within the substation is described in a very straightforward manner. Items which are connected are simply listed and separated by commas, with a full stop at the end of the list. Therefore the entry 'bb1, tr2, cb2, bb3.' states that bb1 is connected to tr2 which is connected to cb2 which is connected to bb3.

Following the connectivity data for each substation is a section of data which describes the association between SCADA message names and network nodes. The data for the Milton substation is:

```
Indication
{
"TRANS 1", cb3 ; "TRANS 1" is cb3
"TRANS 2", cb2 ; "TRANS 2" is cb2
"33 BSECT", cb1 ; "33 BSECT" is cb1
}
```

which specifies the association of the indication message names TRANS 1, TRANS 2 and 33 BSECT with circuit breakers cb3, cb2 and cb1 respectively. This allows the alarm processor to establish, for example, that the message:

```
0924 MILTON 33 BSECT OPEN
```

refers in fact to cb1.

One notable feature of both the topology database and the rules database is that both are read once only during the start-up of the alarm processor. They are processed into an internal form which ensures that the alarm processor can access the data in the quickest most efficient way possible. In other words, the external representation of the data is intended to be readable by humans, but for most convenient access by computer it is actually held in memory in a different form which is more efficient in terms of speed of access and space.

7.5.2 Example of APEX operation

APEX embodies a hypothesizing approach tailored for efficient operation in the alarm processing role, and operates in the manner described by Figure 7.5.

In order to clarify further how this mechanism operates, an example of operation of APEX with a small section of real alarm stream information is presented. Refer also to the network diagram in Figure 7.4.

Sample section of real alarm message stream:

```
0924   MILTON                 TRANS 2  OPEN
0924   MILTON                 33 BSECT  OPEN
0924   MILTON                 AUTOTRIP  OPERATED
0925   WESTFIELD   T1 11KV    OPEN
0925   WESTFIELD   T1/MILT.    OPEN
0925   WESTFIELD   AUTOTRIP    OPERATED
```

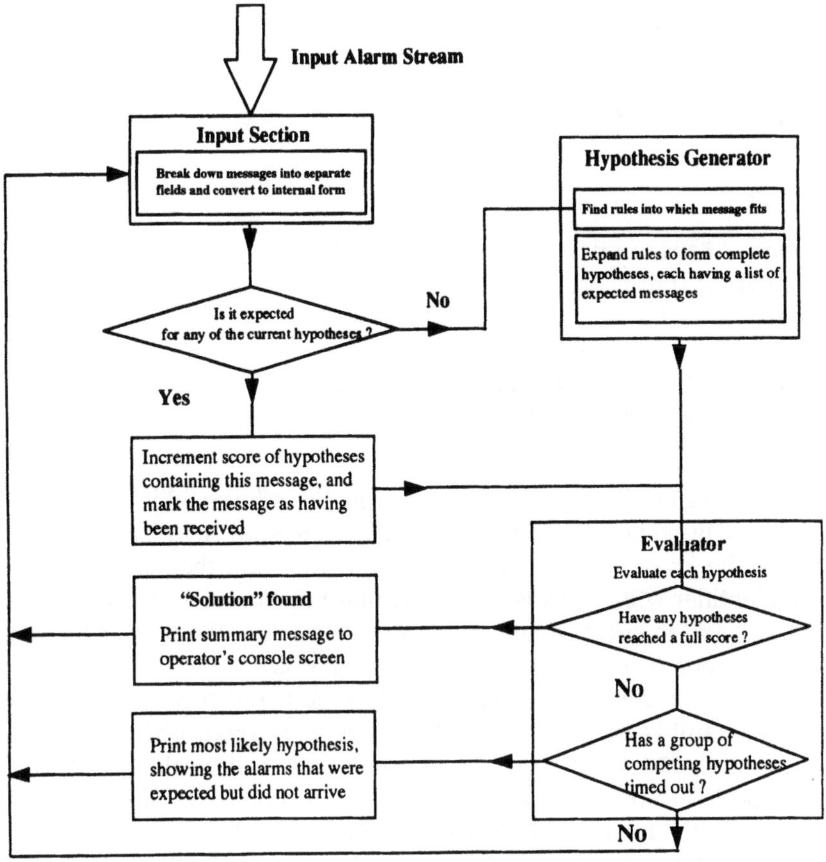

Figure 7.5 Illustration of the hypothesizing approach.

Let us assume as a starting point that there are no current hypotheses held by the system.

On arrival of the first message, `0924 MILTON TRANS 2 OPEN`, there are no hypotheses, therefore a new set of hypotheses must be created to account for this message.

Step 1

The rules database is searched to find all the events in which the opening of a breaker might be expected to occur. The following two rules are thus found:

Rule A:

```
Event "Successful opening of <CB> under operator
  control"
```

```
Expect
{
OpAction OPEN <CB>
Indication OPEN <CB>
}
```

Rule B:

```
Event "Fault on <BlackOut> successfully isolated by
  protection"
Expect
{
Alarm AUTOTRIP OPERATED <StationSet>
Indication OPEN <CBSet>
}
```

The two rules form the skeletons for the hypotheses which are created. The first hypothesis is derived directly from Rule A by simply substituting the known breaker name `Milton cb2` into the rule in place of the identifier `<CB>`. This is a fundamental function of APEX's 'inference engine', and in effect is performing a role similar to PROLOG's inference engine when it instantiates variables with atoms.

Hypotheses 2 and 3 are derived from Rule B and cover the possibility that the fault is on either of the two sides of `Milton cb2`. For hypothesis 2, the `<CBSet>` is derived by finding the set of breakers that would be expected to trip in order to isolate a fault on the `tr2` side of `cb2`. By doing this the related variables `<StationSet>` and `<BlackOut>` become known. Therefore, the `<CBSet>` is `Milton cb2`, `Westfield cb3` and `Westfield cb5`; the `<StationSet>` is `Milton` and `Westfield`; and the `<BlackOut>` is `Milton tr2`, `Milton bb1`, `Milton ln1` and `Westfield tr1`. Hypothesis 3 relates to the possibility that the fault is on the `bb3` side of `cb2`, and hence the `<CBSet>` is `Milton cb2` and `Milton cb4`. This in turn defines the `<StationSet>` to be `Milton`, and the `<BlackOut>` to be `Milton bb3`. Using these known values for `<CBSet>`, `<StationSet>` and `<BlackOut>` the rules are expanded into hypotheses with associated sets of expected alarms.

Step 2

The alarm processor now enters its 'evaluator' phase, where all of the current hypotheses are examined to see if any have reached a full score of expected alarm messages. At this stage, all the hypotheses have a score of 1 because they are all new, and have all received the `Milton cb2 OPEN` message.

Step 3

Each group of competing hypotheses is examined to check that it has not 'timed-out'. A suggested time-out period is 5 min, therefore if a hypothesis within a group of competing hypotheses has not reached a full score within 5 min, the whole competing group is removed, and a message is printed to indicate the most likely hypothesis and the alarms that were expected to arrive but did not. It should be noted that the time-out period could be tuned to suit the telemetry system and would depend on the length of time between a substation's telemetry data being requested by the control centre (the polling period). It would also be valid for the time-out period to be automatically varied depending on current system conditions.

Clearly the current set of hypotheses is new and will not time-out at this point.

Step 4

The alarm processor returns to its input phase to read the next alarm message from the input stream of alarms. The next alarm is:

```
0924 MILTON 33 BSECT OPEN
```

The `33 BSECT` circuit breaker is designated as `cb1` on the network diagram (Figure 7.4), therefore the current message translates to `MILTON cb1 OPEN`.

Step 5

The list of hypotheses is examined to see if the current alarm message is expected for any of the existing hypotheses. Hypothesis 2 does contain the current alarm therefore its score is incremented, and the `MILTON cb1 OPEN` message is marked as having been received. Hypothesis 2 now has a score of 2, whereas the two other hypotheses still have a score of 1.

Step 6

The 'evaluator' phase is initiated again. Still no hypotheses have reached a full score, nor have any timed-out.

Step 7

The processor returns to the input phase to read the next message:

```
0924 MILTON AUTOTRIP OPERATED
```

Step 8

The list of hypotheses is examined to see if the current alarm message is expected for any of the existing hypotheses. This time hypotheses 2 and 3 are those which expect the current alarm and their scores are incremented correspondingly. Hypothesis 2 now has a score of 3, and hypothesis 3 has a score of 2.

Step 9

The 'evaluator' phase is initiated again. Still no hypotheses have reached a full score and none has timed-out.

Step 10

The processor returns to the input phase to read the next message:

`0925 WESTFIELD T1 11KV OPEN`

The `T1 11KV` breaker at Westfield is designated as `cb5` on the network diagram, therefore the message translates as `WESTFIELD cb5 OPEN`.

Step 11

The list of hypotheses is examined to see if the current alarm message is expected for any of the existing hypotheses. This time, only hypothesis 2 expects the current alarm and its score is incremented to 4.

Step 12

The 'evaluator' phase is initiated again. Still no hypotheses have reached a full score, and none has timed-out.

Step 13

The processor returns to the input phase to read the next message:

`0925 WESTFIELD T1/MILT. OPEN`

The breaker on the transformer `T1/Milton` feeder is designated as `cb3`, therefore the message translates as `WESTFIELD cb3 OPEN`.

Step 14

The list of hypotheses is examined to see if the current alarm message is expected for any of the existing hypotheses. Again only hypothesis 2 expects the current alarm so its score is incremented to 5.

Step 15

The 'evaluator' phase is initiated again. Still no hypotheses have reached a full score, and none has timed-out.

Step 16

The processor returns to the input phase to read the next message:

```
0925 WESTFIELD AUTOTRIP OPERATED
```

Step 17

This message is expected for hypothesis 2, therefore its score is incremented to 6.

Step 18

The 'evaluator' phase is initiated again. Now hypothesis 2 has reached its full score of 6. In other words, every one of the six expected alarms has arrived and this makes hypothesis 2 a virtually certain conclusion. The event message for hypothesis 2 is therefore printed out on the operator's console:

```
0924 Fault on (Milton tr2, Milton bb1, Milton ln1,
Westfield tr1) successfully isolated by protection.
```

Step 19

Hypothesis 2 and all its competing hypotheses are removed from the system's working list of hypotheses and it returns to the input phase ready for the arrival of the next message.

For simplicity of explanation the above example of operation only demonstrates the alarm processor handling a small sample alarm stream in which one event is occurring, but the system copes equally well with real recorded alarm streams containing many events occurring simultaneously. Emphasis has been placed on testing the system with such real alarm streams as this is the best way to verify its performance and develop it with a view toward full practical implementation. Further details of APEX can be found in reference 21.

7.5.3 Performance

For a sample real alarm stream consisting of 236 raw alarm messages APEX was able to process the whole stream to produce 48 summary messages in a time of 5 s. This represents a handling rate of 2832 alarms per minute. When one considers that the maximum output rate of alarms from a large energy management system might typically be 500 alarms per minute, it is clear that APEX can cope more than adequately even under such extreme conditions.

7.6 CONCLUSION

The large volumes of alarm messages which potentially can be received at power system control centres cause particular problems for operators such that their ability to maintain a coherent picture of what is occurring on the network is compromised [22,23]. This situation is especially severe during power system emergencies where the need for the operator to maintain control of the system and take the proper corrective action is extremely important in restoring the system safely and efficiently to normal operation. IKBS-based alarm processors have the potential to significantly aid the operator in maintaining his or her control of the system by providing timely information on the events occurring on the network.

Alarm processing is an IKBS application with its own unique characteristics which need to be borne in mind when developing the IKBS. Based on the authors' experience, use of commercially available IKBS shells, or symbolic programming languages such as LISP and Prolog, may have to be precluded at present through their relative lack of performance for real-time systems, but hardware and software improvements are likely to render them more practicable in future. However, there are many applications which are less time critical in nature and permit the use of IKBS shells. At the same time the alarm-processing task is likely to merge with fault diagnosis as the distinctions between the two which are currently imposed by practical constraints are gradually eroded.

An approach to developing a practical IKBS has been described and an example system (APEX) presented. APEX makes use of a hypothesis-based approach to alarm processing and has been developed in the C programming language for reasons of speed, ease of integration and portability.

REFERENCES

1. *Human Factors Review of Electric Power Dispatch Control Centers*, Vols 1–6, EPRI EL-1960, Electric Power Research Institute, Palo Alto, CA.
2. Wollenberg, B.F. (1986) Feasibility study for an energy management system intelligent alarm processor. *IEEE Transactions on Power Systems*, **PWRS-1**(2), 241–47.
3. Amelink, H., Forte, A.M. and Guberman, R.P. (1986) Dispatcher alarm and message processing. *IEEE Transactions on Power Systems*, **PWRS-1**(3), 188–94.

4. Wollenberg, B.F. and Sakaguchi, T. (1987) Artificial intelligence in power system operations. *Proceedings of the IEEE*, **75**(12), 1678–85.
5. Schulte, R.P., Sheble, G.B., Larsen, S.L., Wrubel, J.N. and Wollenberg, B.F. (1987) Artificial intelligence solutions to power system operating problems. *IEEE Transactions on Power Systems*, **PWRS-2**(4), 920–26.
6. Prince, W.R., Wollenberg, B.F. and Bertagnolli, D.B. (1989) Survey on excessive alarms. *IEEE Transactions on Power Systems*, **4**(3), 950–56.
7. Tesch, D.B., Yu, D.C., Fu, L. and Vairavan, K. (1990) A knowledge based alarm processor for an energy management system *IEEE Transactions on Power Systems*, **5**(1), 268–75.
8. Bijoch, R.W., Harris, S.H. and Volkmann, T.L. (1989) Intelligent alarm processor at Northern States Power, in *Proceedings: Second Symposium on Expert Systems Applications in Power Systems*, Seattle, Washington, July, pp. 79–83.
9. Shoop, A.H., Silverman, S. and Ramesh, B. (1989) Consolidated Edison system operation computer control system (SOCCS) alarm advisor SAA, in *Proceedings Second Symposium on Expert Systems Applications in Power Systems*, Seattle, Washington, July, pp. 84–88.
10. Shoop, A.H. and Silverman, S. (1992) A real time alarm processor. *Electrical Power and Energy Systems*, **14**(2/3), 108–13.
11. Pfau-Wagenbauer, M. and Nejdl, W. (1993) Model/heuristic-based alarm processing for power systems. *Artificial Intelligence for Engineering Design Analysis and Manufacturing (AI-EDAM)*, **7**(1), 65–78.
12. Kiernan, L. and Warwick, K. (1993) An adaptive alarm processor for fault diagnosis on power transmission networks. *Intelligent Systems Engineering*, **2**(1), 3–14.
13. Wen, F. and Chang, C.S. (1996) A probabilistic approach to alarm-processing in power systems using a refined genetic algorithm, in *Proceeedings of the International Conference on Intelligent System Applications to Power Systems (ISAP'96)*, pp. 14–19.
14. Talukdar, S.N., Cardozo, E. and Perry, T. (1986) The operator's assistant: an intelligent, expandable program for power system trouble analysis. *IEEE Transactions on Power Systems*, **PWRS-1**(3), 182–87.
15. Cardozo, E. and Talukdar, S.N. (1988) A distributed expert system for fault diagnosis. *IEEE Transactions on Power Systems*, **3**(2), 641–46.
16. Hemmelman, K., Borys, S., Graffy, J., Goeltz, R., Purucker, S. and Tonn, B. (1989) Communications alarm processor, in *Proceedings of the Seminar on Expert Systems Applications For The Electric Power Industry*, 5–8 June, Orlando, Florida.
17. Cheon, S.W., Chang, S.H. and Chung, H.Y. (1993) Application of neural networks to multiple alarm processing and diagnosis in nuclear power plants. *IEEE Transactions on Nuclear Science*, **40**(1), 11–20.
18. Cheon, S.W., Chang, S.H. and Chung, H.Y. (1993) Development strategies of an expert system for multiple alarm processing and diagnosis in nuclear power plants. *IEEE Transactions on Nuclear Science*, **40**(1), 21–30.
19. Corsberg, D. (1986) Extending an object-oriented alarm-filtering system, *Expert Systems In Government Symposium*, October, pp. 80–87.
20. Danchak, M.M. (1988) Alarm messages in process control. *InTech Journal*, May, 43–47.
21. Young, D.J., Lo, K.L. and McDonald, J.R. (1992) Development of a practical expert system for alarm processing. *IEE Proceedings C*, September.
22. Sprenger, W. *et al.* (1996) Compact and operation oriented visualization of complex system states, *CIGRE* 1996, paper 39–107.

23. Vale, Z.A., Faria, L. and Ramos, C. (1996) Towards more intelligent and adaptive user interfaces for control center applications, in *Proceeedings of the International Conference on Intelligent System Applications to Power Systems (ISAP'96)*, pp. 2–6.

8

Event and fault diagnosis

Graeme M. Burt and James R. McDonald

8.1 INTRODUCTION TO THE PROBLEM OF FAULT DIAGNOSIS

The power of supervisory control and data acquisition (SCADA) systems employed in electrical power systems has continuously increased since their initial introduction. Together with the advances in computing power available at system control centres, the effect of this has been a significant increase in the volume of data presented to system operators. In certain instances, control centre staff have failed to keep track of the interpretation of system conditions following faults. There is a recognized need to provide on-line aids to the interpretation of SCADA system data, such as event and fault diagnosis systems. Event and fault diagnosis is closely related to alarm processing, with the distinction lying mainly in the shallower depth to which the analysis is performed for alarm processing.

8.1.1 Events

The fault diagnosis of power systems is primarily concerned with identifying faults on power system equipment. The protection system fulfils a role of monitoring the primary plant, and the SCADA system monitors protection and switchgear activity. By monitoring these 'secondary effects' it is therefore possible to diagnose power system faults, although the task is then complicated by the behaviour of protection, switchgear and SCADA system equipment, as well as that of the power plant. Fault diagnosis thus includes the identification of system faults, protection maloperations, switchgear maloperations and telemetry malfunctions. Protection and switchgear maloperations include failure to operate (i.e. false non-operation), false operation and slow operation. Telemetry malfunctioning relates to lost alarms. These aspects are unpredictable, widespread, very fast, and can lead to complex changes in the network. Fault diagnosis is by no means a trivial task.

8.1.2 Reasons for fault diagnosis

With the task of fault diagnosis sometimes being very involved, it must be clearly justified. The reasons for performing fault diagnoses may be looked at in three time frames: immediate, short term and long term.

Immediately following a disturbance and its isolation, identification of the faulted plant would allow the adoption of appropriate procedures to restore the system to as near normal conditions as possible. Although the system operator need not know the fault position before initiating system restoration, recognizing the faulted plant allows unnecessarily isolated plant to be identified and re-energized by local or remote control. This is particularly relevant where excessive areas of the network have been isolated as a result of protection/switchgear maloperation. By exercising quick and effective control, guided by a fault diagnosis, customers off supply (and hence lost revenue) can be minimized.

With regard to short-term applications, the fault diagnosis will identify protection and switchgear maloperations. Such findings will indicate that detailed investigation of the protection system and maybe switchgear is required; alternatively unnecessary investigations can be avoided.

If properly maintained, records of fault diagnoses would provide a valuable source of statistical data for assessing the long-term performance of plant, protection types and switchgear. This is a long-term investment and must be done conscientiously as faults and maloperations are not very common.

8.1.3 Basis for diagnosis

As switchgear and some protection equipment is wired into utility SCADA systems, its activity is known at control centres; alarmed protection and switchgear activity is therefore the observable evidence which is caused by and which indicates the unobservable system faults.

Sample point measurements of voltage and current are also supplied by SCADA systems, but are not of direct relevance to fault diagnosis. This is in part due to the transducers being typically limited to 125% of full load, and partly due to the significant delay between polling cycles making it unlikely for the during-fault system parameters to be captured.

Another source of valuable data is disturbance recorders [1]. Although SCADA systems can transmit recorded fault summaries, the full complement of data is so voluminous that it must be communicated via a modem with a consequential delay. In addition, microprocessor-based relays can provide current magnitudes measured at set time instants before their operation, and sometimes the fault location can also be estimated. These sources provide useful data which could support the task of fault diagnosis, but only after the unavoidable communications delay.

8.1.4 Reasons for a fault diagnosis system

Traditionally, fault diagnosis has been performed by the operator in the control room, and by protection engineers much after the event. However, utilities are seeking tools to help their operators diagnose faults because of the changing nature of the task and the advantages of such aids. Even back in 1969, Dy Liacco and Kraynak recognized some of the advantages of having an operator aid for analysing system trouble [2].

Utilities have had difficulty striking a balance between supplying more SCADA data to increase knowledge of system conditions especially following disturbances, and restricting the generation of alarms in order to avoid swamping operators with unnecessary information. The dissatisfaction of operators is evident in the results of a survey of utilities [3]: whereas 65 out of 87 respondents declared the operators to be satisfied with alarms presented to them during normal conditions, this figure fell to 21 out of 87 when related to emergency conditions. A fault diagnosis system could increase knowledge of system conditions, but not require the operator to be presented with excessive alarms.

More alarms are now being generated as a result of the developments in SCADA system technology [4], and the computer systems performing SCADA functions have generally been upgraded. These improvements have allowed more equipment to be wired into SCADA systems, and also facilitated the collection of data from adjoining power systems and independent generators. As a result, possible data retrieval during disturbances is now at a rate above that which operators are able to interpret. The additional real-time data should provide more variables in decision making and, together with the extension of telecontrol, more possible courses of corrective action. A system with the required diagnostic ability, and with consistent behaviour, is therefore required to perform fault diagnosis and give the advantages described in section 8.1.2.

Due to a general slow-down in large-scale generation system development, the operational policy in many utilities has been to have reduced reserves. Furthermore, a greater use of existing plant is stressed due to the environmental constraints many utilities have had imposed. A fault diagnosis system would help the operator better manage the utility's available resources by rapidly identifying faults and consequently the minimum part of the network which need remain isolated.

Major disturbances are often the result of relatively minor faults following on from other faults which reduced system security. Whereas automatic control and protection fulfils a role in maintaining system voltage and frequency, e.g. by load shedding, these actions can be initiated too late to prevent a major disturbance. A fault diagnosis system could help in the rapid recognition of the vulnerability of the system, and hence permit pre-emptive control to be taken to avoid more serious consequences, such as the famous New York blackout of 1977 [5].

A fault diagnostic system could also give added advantages to the function of a control centre. For example, the system could be used to automatically generate fault reports of consistent quality, and also maintain fault, operation and maloperation statistics.

8.1.5 IKBS-based fault diagnosis

The developments in utility computer systems have made the industry much more suitable for IKBS implementation. Communication networks are being used by utilities and several new control centre computer systems have distributed architectures. Moreover, the establishment of widely accessible corporate databases means that the necessary network data can be more easily obtained.

IKBS techniques can provide the means of developing a system for the fault diagnosis of power system networks. Due to the knowledge-intensive and symbolic nature of the fault diagnosis task, knowledge based or IKBS techniques are a suitable means of achieving the benefits of an automatic system described previously.

8.2 REVIEW OF IKBSs FOR NETWORK FAULT DIAGNOSIS

The fault diagnosis of power system networks has received a significant amount of interest in recent years, the aim being to develop IKBSs which would be useful in the control centre environment or in normal post-fault utility activities. However, long before any significant interest in IKBSs, the principles of such a fault diagnostic system for power systems were set out in ASTA.

8.2.1 The 'automatic system trouble analysis' system

Automatic system trouble analysis (ASTA) was presented by Dy Liacco and Kraynak [2] as a procedure for identifying the general state of a network and diagnosing 'trouble'. Areas where faults occurred, protection and switchgear failures and false operations are identified as the SCADA alarms are received. Although the authors make no mention of artificial intelligence, the expression of the analysis as discrete logical 'equations', qualifies this system for being classed as the forerunner of IKBSs for power system network fault diagnosis.

The network is represented as a collection of circuits, each circuit being the smallest element which can be automatically isolated from the rest of the network. The breakers that should trip for a fault on each circuit are defined together with their associated primary and backup relays. The analysis is based on the received switchgear and protection alarms, with voltage indications being used to establish the status of generating stations and certain customer installations.

Operating in real time, ASTA is triggered by a primary or backup relay operation or breaker trip alarm, but does not commence analysis until the system is no longer changing. At this point both the final network status and the history of alarms are used, in what amounts to a pattern-matching approach. Faulted circuits are identified as those circuits which are disconnected from all live sources by tripped breakers, the latter having at least one correctly operated main protection relay. Protection and switchgear maloperations are likewise identified by logical rules. The analysis suffers, however, from a requirement for the perfect receipt of all breaker trip alarms, and the need for the developers to set up a large look-up table identifying the breakers that should trip for a fault on every item of plant.

The described analysis equations could have produced a valuable operator aid, although there is no evidence of ASTA actually being realized. However, the establishment of these analysis principles and the identification of the potential advantages, given the date of publication, deserves recognition.

This work spawned the development of many fault diagnostic IKBSs whose diagnosis follows the same strategy of recognizing blackout areas from SCADA activity.

8.2.2 IKBSs based on ASTA ideas

An IKBS was developed in Prolog for the off-line fault diagnosis of electrical power networks by Fukui and Kawakami [6]. Most likely faulted plant, switchgear and protection failures and false operation are identified from a list of tripped circuit breakers and operated relays. The preliminary identification of fault locations is according to the principles of ASTA (i.e. the fault must lie in a blackout area) as is the identification of failures. Particular possible fault positions are identified using related relays' protected areas: faults are located in the intersection of related relays' protected areas. Rules automatically derive the protected areas according to which of five simplified types the relay belongs. Where more than one blackout area exists in the network, multiple faults are assumed and each one diagnosed on the above basis. The 'most possible' fault locations are identified from the alternatives as those which include the least number of failed and false operations. (Claus *et al.* [7] describe an IKBS for the fault diagnosis of power systems which is largely based on the Fukui and Kawakami system.)

Wollenberg and Sakaguchi [8] also describe a fault diagnostic system following the same philosophy. Their system was realized in the production rule language OPS5.

A hybrid IKBS was developed by Girgis and Johns [9] which follows a similar philosophy in identifying faulted plant and breaker failures to that above, and is mainly written in Prolog. Supporting algorithms are included for the classification of network faults and estimation of line fault location. The method of classifying suspected faults is based on voltage and current phasors (the type stored by microprocessor-based relays), and is also described by Watanabe and Oobe [10], who refer to it as a scalar comparison method. The comparison that is central to this approach is that between pre-fault and fault magnitudes. The fault type classified, together with the type and number of phasors available, are the bases on which algorithms are selected for the calculation of the distance to fault.

The IKBS described by Kirschen *et al.* [11], DSF (diagnosis of system faults), differs in that a hypothesize-and-test strategy is used. However, the generated hypotheses are again based on ASTA's recognition of blackout areas. The hypotheses are tested using the telemetered relay operations and have implied relay and breaker maloperations added, before being ranked on the basis of certainty factors. In keeping with the ASTA approach, this system waits for system activity to cease during which no help is given to the operator. This may have been necessary in traditional control centres where the full mainframe processing power would be required for SCADA functions, but it is unnecessary in a modern control centre, since it would be unlikely for the fault diagnosis to be performed on the processor used for SCADA functions.

In addition, Hotta *et al.* [12] describe an IKBS developed as a control centre restoration guide, but which incorporates a limited fault diagnostic function. It was realized in OPS83 and C, and again involves faulty-equipment identification based on blackout areas and the recognition of failed breakers.

8.2.3 A distributed approach

Distributed diagnosis (D2) was developed by Cardozo and Talukdar [13]. Its predecessor was the operator's assistant [14,15] which consisted of a 'blackboard' connecting various problem solvers. The IKBS uses online circuit-breaker alarms in order to identify faulted lines and busbars, temporary faults, breaker failures, failures of timing relays and blocking signals, telemetry malfunctions, and line overloads and out-of-tolerance voltages. A distributed problem-solving approach was adopted, involving a hybrid of non-algorithmic and algorithmic programs realized using OPS5, LISP and C. A 'manager', 'proposal crew', 'construction crew' and 'demolition crew' make up the team of co-operating programs which are run on various processors in parallel.

A nodal representation of the network is again used. The operator's assistant had a significant amount of protection knowledge, and had access to a discrete event simulator which could be used when unsure of a hypothesis.

The diagnosis attempts to identify hypotheses which could account for the observed activity (some telemetry malfunctions can be tolerated). The operator's assistant used a pattern-matching approach to identify valid hypotheses from those it generated. In order to cope with lost telemetry, D2 uses a synthesis approach instead: the distributed programs effectively grow hypotheses by incrementally adding faults and maloperations in order to cover the received alarms. The created hypotheses are ranked according to descending number of equipment maloperations and telemetry malfunctions. However, the main shortcoming of D2 was its excessive use of computing resources: an example of this is the reported 75 s it took per hypothesis when using three Microvaxes.

8.2.4 Fault diagnosis using object-oriented features

The IKBS known as FAUST (fault analysis using simulation and testing) [16,17] also uses a blackboard system with co-operating modules, although it is implemented on only one processor. FAUST was realized using the Poplog language environment (incorporating POP11, Prolog and an object-oriented programming package called Flavours). Its purpose is to diagnose line and transformer faults, temporary faults and stuck breakers on 132 kV and 33 kV distribution networks, and uses mainly switchgear alarms. Moreover, this system addresses many of the issues required of an on-line application. An object-oriented representation of the network is used by the co-operating modules, namely a new hypothesis generator, a simulator, a monitor and a reviewer.

Alarms are firstly compared by FAUST with any existing hypotheses' expected alarms, which are then updated accordingly. Also, on receiving a breaker trip alarm, the adjacent circuit breakers are identified. If a sufficient number of these are also open, then FAUST generates hypotheses based on faults on the lines and transformers in the area bounded by these breakers. These faults are simulated by propagating along the path between hypothesized fault location and the adjacent breakers. When no more alarms are expected, or when a few minutes have passed, FAUST identifies and displays the hypotheses with the fewest number of expected alarms not arrived.

A similar approach to the identification of possible faults is taken by AHFA. AHFA is an IKBS for fault diagnosis described by Esp and Cahill [18] and developed in Smalltalk-80. AHFA uses both protection and switchgear alarms, although protection activity only fulfils a supportive role, and an extensive object-oriented representation of the network. It identifies faulted busbars, lines and transformers according to a similar philosophy to FAUST. Although limited in some aspects, AHFA does have the ability to use autoreclosing breaker and isolator activity in order to identify if the fault is permanent or temporary and to locate faults more accurately.

8.2.5 Localized fault diagnosis

Jeyasurya *et al.* [19] describe an IKBS for implementation in the substation computer of a modern integrated control and protection system. Information on relay operations and breaker trips is used to identify single fault location and failed protection and switchgear. The IKBS was developed in OPS83, and uses the forward-chaining strategy. The diagnosis determines the protected plant of each of the operated relays, and finds the items of plant common to all the relays. These are the possible fault locations. Where local information is inadequate for a diagnosis, the authors suggest that the station computer could obtain other substation data via the intersubstation communications provided by the integrated control and protection system.

Although not physically located at the substation, the diagnosis of a GESTAL application is based on individual substation models. Ares and Girouard [20] describe GESTAL as a tool for building IKBSs for alarm processing and fault diagnosis of power systems. LISP and ART were used to develop GESTAL, but it has also been implemented using C and OPS83. GESTAL has a frame-based model for each substation and uses switchgear and protection alarms to identify possible fault locations, failed equipment, relationships between multiple faults and the sequence of operations. The substation models essentially describe the characteristics and behaviour of the protection and recovery systems in a simple structured language.

A GESTAL-based IKBS uses a focused data-driven approach. The data-driven inference basically grows directed graphs in accordance with received alarms with causal/associative arcs linking alarms. While obsolete alarms and inconsistent diagnostic graphs are removed, accepted and completed diagnostic graphs are translated to a natural language form for presentation. Ares and Girouard have demonstrated the validity of a GESTAL-based IKBS for real-time fault diagnosis, reporting the generation of some diagnoses in under 1 s.

8.2.6 Applications where SCADA data is lacking

Balakrishnan and Pahwa [21] propose a computer-assisted intelligent storm outage evaluator (CAISOE) for distribution systems. That part of the distribution system which uses fuses is also included in the diagnosis. It was implemented using a LISP-based tool. CAISOE can locate multiple faults and classify them as high-impedance or short-circuit and broken- or unbroken-conductor faults. CAISOE does not use SCADA alarms, but instead requires the placement of voltage sensors at the location of breakers, fuses, tap-off points, and after small groups of distribution transformers. A dedicated radio channel is required to communicate a sensor change of state (a 'symptom') to the control centre. The intelligent symptom processor suppresses secondary symptoms according to a semantic network representation of the network topology. Abstract fault locations are generated from the primary symptoms, and are refined into actual faults by symptom–fault production rules specific to the abstract fault locations. Clearly such a system could only be considered where a SCADA system is not in use at that voltage level.

Though less automatic, the IKBS described by Hsu *et al.* [22] does not require extensive modifications to the existing distribution system, nor additional information which is not available. This IKBS was realized in Prolog for locating faults on distribution systems between primary feeders and customers. Faults on this plant are cleared by fuses; no knowledge of the isolated plant is automatically available at the control centre. The developers were faced with the further problem that there was no accurate schematic of the network in the control centre for which this IKBS was developed. The network data it does have includes a nodal feeder circuit configuration representation, plant location, location and telephone numbers of key customers (those with technicians responsible for the electricity supply), and plant with a higher risk of failure.

The diagnosis of Hsu's system is based on calls from customers whose supply has failed. Following such calls, the IKBS identifies those branching points in the vicinity of each customer as possible supply routes from primary feeders. The intersection of customers' possible supply routes is taken as containing the likely fault location. The telephone numbers of key customers supplied from branching points on these routes are displayed to the operator so that he may contact the technician on these premises to ascertain if the supply has been lost there too. The resulting information, together with the knowledge of higher risk plant, is used to narrow down the possible fault locations.

8.2.7 A system suited to parallel implementation

Okamoto *et al.* [23] describe an IKBS which holds the promise of effective implementation in a parallel-processing environment. It identifies fault sections (single and multiple faults), protection and switchgear maloperations and telemetry malfunctions. An object-oriented representation of the power system (in a LISP environment) is involved, but is not used directly in the fault diagnosis. It is used to automatically derive a semantic ('cause–effect') network every time the power system is reconfigured. The semantic network links fault section, relay and breaker nodes by the following arcs: protected by (main protection), causes tripping of and (breaker) failure causes operation of (backup protection). Protection and switchgear alarms and the semantic network are used in the diagnostic process which is suited to parallel processing.

The diagnostic process commences with the identification of protected fault sections from the operated protection relays, and the association of each tripped breaker with its controlling protection relay. For each fault section hypothesis, protection and switchgear maloperations and telemetry malfunctions are identified by recognizing patterns in linked nodes. Multiple fault hypotheses are created if false relay activity has been identified. The most probable hypotheses are those with the lowest combined total of system faults, maloperations and malfunctions. While there are clear processing advantages in using the described technique, its main failing is the processing time required to automatically derive the semantic network (6 s is reported for a very simple example).

8.2.8 A system for distribution feeders

An IKBS is reported to have been under development for the fault diagnosis of 10–20 kV distribution feeders and also to help reconfigure the network. Partanen *et al.* [24] used the frames, rules, object-oriented programming and LISP provided by the GoldWorks toolkit. The diagnosis is based on data available from the SCADA system and includes switchgear and relay alarms from the primary substations, status of fault current detectors on the distribution lines, and weather data. It also requires the distance to fault data from the operated relay.

Following a breaker trip alarm from a primary substation, the diagnosis is initiated on the basis that the breaker's feeder is faulty. Using the status of the fault detectors, the possibly faulted line sections are identified. The most likely fault location is then that section whose electrical distance from the primary feeder matches the relay's calculated distance to fault. The nearest upstream isolator is then advised to be closed. The system, however, does not allow for failure of protection/switchgear equipment.

8.2.9 CRAFT – customer restoration and fault testing

Customer restoration and fault testing (CRAFT) is an IKBS whose purpose is to identify the faulted section of multiply tapped 115 kV transmission lines and propose restoration actions [25]. It follows from work by Tomsovic *et al.* [26] and has been operational in a control centre since mid-1988. CRAFT is production rule based and is realized in OPS83. The basis of its diagnosis is telemetered switchgear alarms and knowledge of the function and time settings of automatic switchgear.

When a non-control alarm is received, CRAFT considers that switchgear's line to be faulty. The diagnosis is delayed for a period of 1 min to allow for the completion of the system's response to the fault, during which other alarms may be processed. The isolator status validation knowledge is then applied to correct the isolator status data. The identification of the faulted section is then achieved by the application of empirical rules which use the post-fault configuration of the switchgear on the line. Empirical rules also advise some control actions.

Although automatic isolator alarm malfunctions are acknowledged, CRAFT assumes that all breaker alarms are correctly received and that switchgear never maloperates. However, CRAFT is a valuable, though limited, tool for appropriate utilities which has been successfully implemented in a control centre.

8.2.10 SEPT

Fauquembergue and Esquevin [27] describe SEPT, an IKBS for diagnosing incidents on a power system. SEPT has been operational in an Electricité de France (EDF) area control centre since 1990. The most likely faulted plant and protection maloperations are identified in an on-line role. The IKBS uses event-recorder data from one substation as the basis of its diagnosis. Data is automatically retrieved via a PC and modem following a protection operation being identified in the control centre remote terminal unit (RTU) information. Knowledge is represented by production rules, objects and methods, and includes a deep protection system model.

The diagnosis consists of four parts. The 'verification of the protection systems' uses a pattern-matching approach to verify the dynamic activity of the protection or identify anomalies. Heuristic rules and interval filtering are used in the 'search for elementary incidents' which seeks to divide the event recording into intervals of time containing activity relating to individual incidents. The 'fault section estimation' considers each high-voltage item of plant to be possibly faulty, for which each verified protection relay (no anomalies) can vote for or against on the basis of the protection system model; the section with the highest vote is defined to be faulty. The 'synthesis' uses heuristic rules to perform a final check, to use the faulted section conclusion to identify further protection maloperations, and to distinguish between temporary and permanent faults. If one of the diagnosis parts fails then SEPT will consult the operator.

SEPT has been proved to be a practical tool for a utility, and demonstrates the use of substation sequence of event recorder data. However, it is not particularly suited to system-wide application since its knowledge is very much specific for a particular substation arrangement and standardized protection system.

8.2.11 An IKBS developed by Amantegui *et al.*

Amantegui *et al.* [28] describe an IKBS for fault diagnosis in power systems. An initial prototype was developed using a shell (M1), C and assembly language, and was operational for one-and-a-half years. The final system was, however, developed using ART and C, and involves rule-based parts, object-oriented parts and procedures. The IKBS uses SCADA alarms.

The fault diagnosis reasoning is performed by five separate modules. The 'problem-solving kernel' controls the performance of the IKBS. The shallow knowledge which forms the 'selector of the possible fault locations' generates hypotheses of the fault location with associated certainty factors. The 'hypothesis validation and fault location' module identifies the alarms related to each hypothesis and adds deviations where necessitated by the received alarms. The hypothesis which includes the fewest maloperations is selected as the conclusion. If a conclusion is not reached, then the 'simultaneous disturbance analyser' attempts to generate multiple fault hypotheses by trying to recognize geographically distinguishable clusters of alarms, each relating to one fault. Among the roles of the 'time pursuit of a disturbance' module, is the validation of hypotheses in line with automatic resetting/reclosing and operator action.

Although the IKBS provides a very good fault diagnostic role, it appears only to make use of the normal network arrangement instead of the present configuration.

8.2.12 Further comments

In addition to the above systems, other closely related work has been performed. Keronen [29,30] is developing an IKBS for operation planning and event analysis, called EKA. It has been developed using KEE, LISP and Fortran, and contains a comprehensive hierarchical frame-based representation of the power system. The event analysis module is said to include causal, eventual and temporal relations together with pattern matching and simulation-based reasoning.

Kaminaris *et al.* [31] developed a rule-based IKBS called SUBES for the related task of substation troubleshooting. Developed in Turbo Prolog, the IKBS uses a tree structure which is traversed as the diagnosis progresses. The reasoning may also advise tests as well as generating maintenance schedules. Protopapas *et al.* [32] described a similar IKBS.

Certain functions have been provided together with the core fault diagnosis role of the above systems. These include alarm processing/filtering (D2 and FAUST) and live interactive graphics (AHFA and EKA). However, it is clear that many of these IKBSs have certain deficiencies. These include unrealistic assumptions, vast computer resources requirements, dependency on having complete information (i.e. their diagnosis is not robust), indecisiveness (i.e. inability to discriminate between several possible diagnoses) and inappropriate methodologies for real-time implementation.

8.3 THE DEVELOPMENT OF AN IKBS FOR FAULT DIAGNOSIS: RESPONDD

8.3.1 Introduction

An IKBS called RESPONDD has been built by the authors for the fault diagnosis of power systems [33]. It is believed that the system has been built in such a way that many of the aforementioned shortcomings in other systems have been avoided. The early developments in the fault diagnostic IKBS were tested on a small hypothetical transmission/distribution network. However, the IKBS was later applied to a utility's transmission/subtransmission system. The portability of the core diagnostic knowledge was demonstrated when the IKBS was also applied to the main electrical power network of a nuclear power station: little change was required in the central diagnosis module.

Recent developments in power systems have indicated that IKBS techniques could provide systems that are both suitable and necessary for on-line network fault diagnosis [34–36]. However, fault diagnosis of power networks differs in several respects from those problem-solving tasks for which general (or 'first generation') expert systems were developed. These differences had significant bearing on the design of the developed IKBS and include the following.

1. Automatically supplied evidence: the SCADA system and not a user is the main source of the information on which the diagnosis is based. Data retrieval is thus automatic, fast, relatively random and does not require user dialogue.

2. Size of the object of diagnosis: the power system network is very large, and the number of possible conditions, diagnostic conclusions and possible configurations is vast. (For example, the ScottishPower transmission network contains 128 substations with 5000 double digital indications and 18 000 single digital alarms in total.) Abstract networks, like semantic networks or scripts, were considered to be unsuitably large.
3. Depth of reasoning: since knowledge is equally appropriate in various parts of the system and a robust diagnosis is desired, a significant depth of reasoning and system modelling is required.
4. Legitimacy of many answers: the process of fault diagnosis can reasonably generate several valid answers. The reasons for this shall be detailed in section 8.3.5. The means of producing several alternatives rather than one possibility must be provided.
5. Interfacing to numerical algorithms: the ability to interface with standard power application software would be a valuable attribute. Although not an immediate necessity, this facility would extend the system's potential benefits.

In the light of the above, Prolog (specifically Quintus Prolog [37]) was chosen as the means of developing the fault diagnostic IKBS. In the following description of the IKBS development, a different typeface is used for Prolog terms.

8.3.2 Structure of the diagnosis: its results

In a generation system, the plant is more localized and strongly interconnected than in a transmission or distribution system. There is seldom more than one simultaneous (i.e. close in time) disturbance. However, in a transmission or distribution system, the volume, dispersion and vulnerability of plant is such that there may be several on-going independent disturbances at any point in time. There could be several alternative explanations attempting to describe any disturbance. Moreover, since the fault diagnosis is performed on-line, the diagnostic conclusions are subject to change. As a result, the diagnostic conclusions cannot only be expressed as textual output, but must be held in the knowledge base as knowledge.

Diagnostic conclusions are formed according to a three-tier hierarchy. This provides for the separation of unrelated disturbances, the comparison of competing conclusions and the combination of related occurrences. The three levels which each diagnostic conclusion has are:

1. the disturbance
2. the competing scenario
3. the causal event.

Potentially related behavioural activity belongs to the same disturbance. Each disturbance has an area of involvement which identifies that part of the network which could be involved in the same disturbance due to its connective proximity. Initially it was thought that a planar grid could be used to divide the network into disturbance areas. Along the boundaries of these areas, protection on one side of the boundary would not be affected by system events on the other. However, such boundaries could only exist where there were unprotected parts of the network, and protections overlap for this very reason of avoiding unprotected areas in the network. A planar grid is therefore inappropriate. While an array of overlapping areas could be established, the high degree of protection overlapping means that any disturbance area would consist of several defined areas. For this reason, the technique adopted was to form the disturbance area during the diagnosis directly from the topology. The actual reasoning used to form the area is dependent on the diagnostic philosophy used. A ranked table of competing scenarios is maintained for each disturbance.

Competing scenarios are alternative explanations of the same disturbance. Each scenario may contain several causal events. Causal events include system faults, failed equipment, false activity and telemetry malfunctions. A system fault includes the faulted item of plant, the generic type of fault (ground or phase fault), and the fault permanence ('permanent' or 'temporary' fault). Failed equipment is particularly protection and/or switchgear which has not operated, while the phrase false activity is applied specifically to devices which have wrongly been active when they should have remained inactive. Failed equipment includes both the failure to detect a fault and the failure to operate following a control signal from a relay. Missing alarms result from telemetry malfunctions, and relate to activity which has taken place but has not been alarmed. The telemetry which supports the causal events is also expressed in the scenario fact. Each scenario fact has nine arguments in total: disturbance identifier, ranking position, measure of uncertainty, system faults, device failures, supportive alarms, false activity, predicted/missing alarms and an argument for extra knowledge which relates alarms and failures to specific causal events. The following example illustrates this.

```
scenario(
disturbance_id:   -19999,
rank:             1,
uncertainty:         0.001,
faults:           [phase fault_on ss1 to ss2::l1
                    which_is
                  (permanent or temporary)],
```

```
failures:          [],
supportive:           [operated(rl at ss1), operated(trl
  at ss1), operated(rl at ss2),      operated(trl at
  ss2), tripped(x1 at ss1), tripped(x2 at ss2)],
false:             [],
predicted:           [],
extra:             [events_alarms(details(1990,1,11,15,
                   41,37.0),
                   phase fault_on ss1 to ss2::l1 which_is
  (permanent or temporary),
                   [operated(rl at ss1), operated(trl at
                   ss1),
                   operated(rl at ss2), operated(trl at
                   ss2), tripped(x1 at ss1), tripped(x2
                   at ss2)], [], isolated)]
).
```

8.3.3 Structure of the diagnosis: its mechanics

In the fault diagnosis of transmission system networks, there are two competing factors, namely scale and detail. The scale factor requires the IKBS to be able to diagnose faults throughout the network; the detail factor dictates that the diagnosis should focus down to as much detail as permitted by the available data. Having the diagnostic knowledge at different levels and using a 'focusing' approach resolves this conflict. Consequently a three-tier hierarchical structure involving a 'gut reaction', 'central diagnosis' and 'deeper diagnosis' was developed. In addition to these, performance monitor and management modules have been created.

1. Gut reaction: the top level diagnosis establishes a 'gut reaction' or first impression based on the interpreted alarm. The diagnostic knowledge can identify activity which can only be as a result of manual control. Additionally, automatic activity alarms may be placed in the context of the state of the network by establishing the alarm's relationship to disturbances. The alarm type and nature are determined, with some simple abnormalities being checked for. The knowledge is represented by production rules for clarity; a supplementary inference engine was developed in Prolog in order to use knowledge in this form. The picture of the alarm which is built up is then used to guide the knowledge contained in the central diagnosis.
2. Central diagnosis: the central diagnosis performs most of the diagnostic work, and creates and modifies scenarios. System faults, protection and switchgear maloperations and telemetry malfunctions

are inferred at this level. Differing philosophies have been developed to establish this diagnostic ability, and these have been realized using the standard Prolog logical rule structure. Though less understandable than the gut reaction production rules, these logical rules have realized a depth of knowledge which could not have been effectively represented in such production rules. Despite this, the knowledge remains reasonably understandable due to Prolog's semantics.

3. Deeper diagnosis: the deeper diagnosis makes use of the ranked scenarios produced by the central diagnosis. The deeper diagnosis provides for the use of deeper relationships in system behaviour, non-electrical knowledge, deeper plant models, and will have the knowledge to validate and add detail to previous diagnoses using data from disturbance recorders and microprocessor-based relays.
4. Performance monitor: as well as performing a fault diagnosis, the IKBS monitors the performance of the protection and switchgear. The performance monitor records failed and correctly operated protection or switchgear equipment which is included in the top few concluded scenarios.
5. Management: since the fault diagnosis is a continuous process, the reasoning will not reach a complete conclusion and stop. Therefore, the IKBS has the knowledge to recognize when a system event or a disturbance is complete and can be taken out of consideration. The deeper diagnosis can then be applied to the central diagnosis' final ranked table of conclusions.

8.3.4 An initial prototype: wait-and-see

Faced with the appearance of protection and switchgear alarms, a power system operator may wait until he or she has established that an area of the network has been isolated, before attempting to identify possible faults which caused the creation of the blackout area. The operator then seeks a system fault inside the blackout area which would have caused the operation of those relays which tripped the breakers surrounding the blackout area. This is the philosophy used in the developed wait-and-see fault diagnosis. The on-line SCADA system alarms describing protection and switchgear activity are used, though the switchgear activity is of primary importance. A remote blackout area forms a disturbance area, as do related adjacent blackout areas.

(a) Gut reaction

The 'gut reaction' performs some general checks on the interpreted alarm. The validity of the alarm is tested by verifying that the alarm has not already been instantiated in the knowledge base. The primary/

secondary nature of the alarm is ascertained and, if secondary, the alarm is checked to determine if it was controlled by an operated device and hence expected.

The gut reaction also has rules which relate specifically to the wait-and-see strategy. Since the wait-and-see is primarily circuit-breaker-activity driven, the reasoning following such is greater. Following a breaker trip alarm, the status of the breaker terminals (i.e. electrically dead or live) is established by topologically searching for connected power infeeds. A power infeed is plant which could feed a system fault, and therefore includes machines and supply points (plant connected to undefined parts of the network which can supply power).

If the status of the breaker terminals is live on one side and dead on the other, then a new blackout area has been created. A gut reaction rule compares the breakers isolating the new blackout area with those of existing blackout areas in order to identify adjacent blackouts. Thereby, it is inferred whether it is a new disturbance area or belongs to an old one.

(b) Central diagnosis

Provided that the gut reaction did not indicate the alarm to be unreasonable, the action described by the alarm updates the representation of the network held in the knowledge base. The rest of the response of the central diagnosis depends on the type of the alarm.

Following a protection relay operation alarm, the relay is classed as slow if all its current transformers (CTs) are in a blackout area and it was itself classified as a failure. A false relay has all its CTs on dead plant but was not included in any scenarios. The disturbance scenarios are updated accordingly. If a CT is found on live plant, the possible faults which could have caused the relay activity are identified and asserted in the knowledge base, and the alarm recorded as 'unaccounted for'.

The reasoning following a circuit-breaker trip is dictated by the status of the breaker terminals.

1. Live/live: the breaker activity is logged as unaccounted for, and no more reasoning is performed.
2. Dead/dead: if a controlling relay is known to have been slow then no more reasoning is performed; otherwise, the scenarios are updated with the breaker activity classed as slow if it was formerly a failure, or false.
3. Live/dead: a system fault is presumed to have occurred and been cleared, and the diagnosis continues into more detail to determine the possible system faults.

Related operated relays – i.e. those which could have seen the fault in the blackout area – are then identified. The possible faults are then those faults which could have been detected by all the related primary relays and which are located in the blackout area. Each possible fault is the basis for a temporary single fault scenario.

From the equipment activity list, and using relay auxiliary switch logic, trip relays and switchgear which failed to operate/trip are identified. These failures are independent of the fault, and so are added to each new temporary scenario. In addition, for each possible fault scenario, unoperated protection relays which should have operated are included as failures.

If the temporary single fault scenarios belong to a new disturbance, they are copied as complete single fault scenarios. Otherwise all the temporary scenarios are combined with all the disturbance's existing scenarios.

The first way of combining a new scenario with an old one is on the basis that the events in both scenarios genuinely occurred: the faults, maloperations and corresponding alarms in the two scenarios are added to form a complete scenario. In addition, due to the vicinity of the isolated regions, it is feasible for a blackout area to have been falsely created because a nearby fault destabilized protection: one scenario has its details copied to the complete scenario, while the other has its secondary activity details copied and the primary supportive alarms included as false activity.

(c) Assessment

The wait-and-see IKBS was extensively tested. In terms of analysis timing, the IKBS was acceptable since diagnoses were complete in under 1 min. The diagnosis was relatively decisive since only fault positions inside a blackout area were considered. It was, however, lacking in robustness: success of the diagnosis depended on every switchgear alarm being correctly received. This level of 'brittleness' was unacceptable, and resulted in the development of the hypothesizer IKBS.

8.3.5 Hypothesizer strategy

An operator faced with protection and switchgear alarms may consider it unreasonable to wait for the recognition of an isolated area of the network, especially when he is aware of the possibility of alarms being lost. He would determine which system faults a relay can detect following the receipt of an alarm describing that protection's operation. For each possibility (or 'hypothesis'), he resolves the alarms which he would expect to see if the suspected fault was genuine. By checking whether these expected alarms have been received, hypotheses are discarded or kept. Subsequent activity would influence his confidence in these hypotheses. This philosophy is based on the generate-and-test paradigm and is the basis of the developed hypothesizer IKBS. Again, on-line switchgear and protection alarms are used. However, the hypothesizer philosophy is primarily protection-activity driven. It involves the prediction of activity, and hence has the potential to cope with missing data. A disturbance area, according to the hypothesizer strategy, is the network in the vicinity (by electrical topology, not geographical proximity) of the substation which generated the original protection alarm.

8.3.6 Hypothesizer gut reaction

The hypothesizer 'gut reaction' checks for the same general judgements as does the wait-and-see gut reaction. There are also rules specific to the hypothesizing philosophy. Unlike the wait-and-see philosophy, the hypothesizer suspects a fault following one alarm. Therefore the disturbance area is created, not after the isolation of an area of the network, but following an alarmed relay operation or circuit-breaker trip. The disturbance area is defined by a list of substations. If an automatic activity alarm is from a substation inside a disturbance area, then a rule records the judgement that the activity belongs to that disturbance. Otherwise, a new disturbance area is created.

A new disturbance area of involvement is generated by forming a list of substations from the alarm 'source' substation and those found a prescribed number of substations away which are not already in an existing disturbance area. The disturbance area radius must cater for the distance from the protection's substation to the fault, and the distance on the other side to another non-unit protection since the fault can be fed from both sides in an interconnected network. Therefore the production rules which set the radius make a distinction between unit and non-unit protection relay operations.

8.3.7 Evolution of hypothesizing strategies for central diagnosis

Initially, the central diagnosis had a fixed maximum number of simultaneous hypothesized failures. Following an alarm, power system fault hypotheses were first generated and simulated. Subhypotheses were generated by adding one failure to the successful no-failure hypotheses, and then simulated. These were used to generate more subhypotheses, and the process of adding failures repeated until the maximum was reached. Scenarios had subsequent testing performed on them following the receipt of more alarms: subsequent related alarms strengthened the existing scenarios which predicted the activity. Also, new hypotheses were generated at every opportunity. Since any alarm can be caused by numerous fault-plus-failure combinations, a multitude of hypotheses could be created. A strategy to contain this combinatorial avalanche was therefore evolved.

The strategy was therefore changed so that no failures were initially considered. Only following the receipt of an unexpected alarm was the permissible number of hypothesized failures incremented and new disturbance hypotheses created. However, numerous scenarios were still generated following an unexpected alarm.

Therefore, in addition to being able to generate and subsequently test scenarios, the ability to derive hypotheses was introduced. When an unexpected alarm is received, the IKBS seeks first to develop existing hypotheses by the addition of a failure before seeking to generate new hypotheses. The candidates for failures are those devices with predicted activity in the existing scenarios. This resultant strategy exhibited an acceptably low level of indecisiveness without unacceptably affecting robustness.

(a) Original hypothesis and scenario generation

When the gut reaction has judged an alarm to belong to a new disturbance, then the knowledge in the hypothesis and scenario generation rules is used. The rules are also used following an alarm in an old disturbance if old scenarios remain after the subsequent testing and hypothesis derivation rules have been invoked.

The hypothesizer is primarily driven by alarmed protection relay operations. (However, because such an alarm may be lost and a secondary activity alarm initiates the creation of a new disturbance area, knowledge is included to derive the operated protection relay.) Protection modelling rules determine the possible faults which the operated protection relay could have detected – all other protection relays are ignored at this stage. A system fault is taken from the list as a hypothesis. Before the fully formed hypothesis is qualitatively simulated, the knowledge base representation of the network is formed into the configuration which existed before the first related alarm. The hypothesis is qualitatively simulated as described in section 8.3.10. The possible consequences of the simulation are threefold (this is the immediate, as distinct from the subsequent, testing of hypotheses).

1. Fault not isolated: the genuine isolation of the faulted plant involved switchgear in the undefined part of the network. The reasoning simply backtracks to find the next hypothesis.
2. Genuine activity not included: protection nearer the fault or of higher priority has cleared the simulated fault. Relays suspected of this are identified as profitable hypothesized failures and may be used for derived hypothesis generation. Again, the next hypothesis is sought.

3. Genuine activity included: a successful simulation causes the reasoning to proceed. The activity involved in the simulation is classified as supportive or predicted depending on whether a corresponding alarm has been received or not. A partial scenario is then formed from the classified simulated activity, the hypothesis and any extra information obtained from the simulation.

If they relate to a new disturbance, the partial scenario details are copied into a complete scenario fact which is stored in the knowledge base. Partial scenarios for old disturbances have the outstanding old scenario's details added before being asserted as complete scenarios: for each partial scenario, a new scenario is generated for every remaining old scenario. Backtracking is then started in order to find another hypothesized fault. When all such hypotheses have been simulated, the hypothesis that the operated device did so falsely is dealt with in the same way. In addition, if the permissible failures limit is non-zero, a successful single-fault partial scenario may have a failure added to generate subhypotheses.

(b) Subsequent testing

Subsequent testing updates scenarios following alarms subsequent to the alarm which invoked the generation of the scenario. When an alarm belonging to an old disturbance is received, then all that disturbance's existing scenarios are tested. The four tests shown below are attempted in turn until one succeeds or all fail, for each scenario.

1. Activity was included as predicted: the scenario is replaced with an equivalent new one, except for the reclassification of the newly observed activity as supportive.
2. Active device was supposed failed: since the inclusion of failures is suppressed until there is evidence which is thought to support it, the scenario is not removed as invalidated. It has the latest activity included as slow and its secondary activity included as predicted.
3. Activity was included as possible: the scenario has the group of possible activity removed, the received activity classed as supportive and the other activity in the group included as predicted.
4. Activity was alternative to default activity: in the light of the most recent activity, the old scenario is removed and regenerated, including the newly received activity rather than the default.

(c) Derived hypothesis and scenario generation

Hypotheses are not only generated originally from the relay protection model. As has been mentioned, a scenario may need to be recreated if the alternative to default activity is observed. Having identified a scenario for which alternative activity has occurred, and the actual fault and failure hypothesis to which it pertains, the hypothesis is resimulated. The simulation automatically uses the knowledge of the latest alarm so that the scenario created does not involve exactly the same (wrong) alarms.

Secondly, a hypothesis may be derived on the basis that the latest alarm is evidence of a device failure. The derived hypothesis is formed by adding a failure to a hypothesis which is found in an old scenario remaining after the subsequent testing and has a hypothesized fault which can be detected by the newly operated protection relay. The disturbance's maximum number of simultaneous failures is incremented. For every device with predicted activity in the old scenario, a hypothesis and hence scenario are generated with that device as the failure.

Finally, a hypothesis may be derived based on the failure of a previous qualitative simulation. In certain cases the reasoning may fail to generate any partial, and hence complete, scenarios. This arises when a backup operation alarm has been received and the necessary number of failures has not been permitted. Those qualitative simulations which did not account for the genuine activity will have generated facts detailing those relays and circuit breakers whose correct behaviour could have been the reason that the genuine activity was not simulated. The simulated hypotheses are modified by having these devices, if they have not yet operated, included as failures. The disturbance's number of simultaneous failures is increased by one. Partial and complete scenarios are attempted as before.

8.3.8 Dealing with uncertainty

(a) Sources of uncertainty

Dealing with uncertainty plays a significant role in many problem-solving areas, including the on-line fault diagnosis of power system networks. Although the diagnosis of a disturbance may fail to generate only one possible conclusion, this is not an inadequacy of the diagnosis. There are several reasons why there can be legitimately more than one possible answer.

1. Lost alarms: telemetry system malfunctioning can cause an alarm to be lost. But a failed device will likewise result in no alarm being received.
2. Limits of SCADA system: most, if not all, utilities have not wired all relays into their SCADA systems. The evidence is therefore incomplete.

3. Protection relay characteristics: an operated relay may have detected one of various fault types and locations.
4. Protection relay failure: relays can fail, and it can be difficult to distinguish between correct and incorrect non-operation when no alarm is received in either case.

(b) Problem factors

Various techniques for dealing with uncertainty are well known [38]. These include subjective probability theory, Dempster–Shafer theory, possibility theory, the theory of certainty factors, the theory of likelihood ratios, scoring techniques, and the theory of endorsements. Fuzzy mathematics is also used to deal with uncertainty [39]. The amount of power network, protection and switchgear equipment involved in power systems and the diversity of its behaviour means that the volume of data required by any technique for dealing with uncertainty is significant. The diversity of equipment and the diverse ways in which it is configured, restrict the use of simplifying generalities which could reduce the volume of necessary data. Those techniques which require a complete definition of all possible conclusions (like the Dempster–Shafer theory) are invalidated by the colossal number of possible diagnostic conclusions.

Furthermore, sound statistics are not available. The quality of statistics produced on the basis of operator logs is questionable due to the many other duties an operator has to perform following a fault. Moreover the ability to accurately set subjective values is uncommon due to the rarity of fault events and the small number of large power systems. In addition, the contrast between the assumptions on which these techniques are based and the real behaviour of power and protection systems would indicate that the selection of the techniques on the merit of their mathematical soundness would be unjustifiable. At the same time it was felt that scoring techniques did not make full use of available knowledge, such as the difference in likelihoods of different maloperations. It was therefore decided to develop a compromise technique.

(c) The developed measure – relative likelihood

The absolute value of the uncertainty measure is not the critical feature, rather it is the value in comparison with values for competing aspects. In addition, the values are not calculated from statistics, and the confidence in their validity is not sufficient to merit being called probability. Hence the term 'relative likelihood'. A value range of 0 to 1 is used. Relative likelihoods are set for possible aspects of scenarios (e.g. system faults, predicted alarms etc.), and are combined to form values for specific scenarios. Comparison is only required between coexisting scenarios, and hence between coexisting scenario aspects. Since every aspect of a scenario influences its total likelihood, relative likelihoods are combined by the product rule, as are independent probabilities. For each aspect of a scenario, a typical value may be combined with (i.e. multiplied by) a particular value. The inclusion of a particular value is a means of including knowledge of plant or equipment which is, for example, less reliable than other similar plant or equipment. A learning influence can thus be incorporated.

The technique is not mathematically sound, but does give a reasonable comparative indicator. Each scenario is ranked in a table of competing scenarios according to its relative likelihood. The scenario with the highest relative likelihood occupies the top-ranking position, and is the scenario thought most likely to describe the genuine system state. Relative likelihood therefore provides the means of restricting the number of possible answers which need to be shown to the engineer.

8.3.9 Network data requirements

This IKBS, unlike some, maintains a clear distinction between the network data and the network and diagnostic knowledge. This separation increases the system's portability to other power systems and allows it to be more easily maintained in line with modifications to the power system being diagnosed. In order to make use of the network behavioural knowledge, details are required of the system to be diagnosed. The network data can be broadly categorized as dynamic and static data.

(a) Dynamic data requirements

Clearly, on-line fault diagnosis is based on dynamically changing data which describes, in some way, the behaviour of the system to be diagnosed. This IKBS uses supervisory control and data acquisition (SCADA) system alarms of protection and switchgear activity. The alarms are automatically accessed and are both treated as symptoms of the fault to be diagnosed and used to keep the network representation up-to-date. A typical alarm in the power system on which the development was based might be:

```
13:45:13:00 DUMB2 COAT2 FIRST MAIN PROT OPTD ON
```

The fields in this alarm relate respectively to the outstation time tag, the source substation (`dumb2`), the circuit or feeder (identified by remote end substation, `coat2`), a legend identifying the operation of that circuit's first main relay, and the new status of the relay (on).

(b) Static data requirements

The static data requirements can be very extensive for typically large transmission systems. The data incorporated covers a description of power system plant and of protection and switchgear equipment. The existence of items of plant, switchgear, protection and trip relays, and current transformers (CTs) is detailed in terms of the equipment types for which modelling knowledge exists. Relational data is included which relates items of plant and switchgear (i.e. topology), relates relays and CTs, and relates relays with each other and with switchgear (i.e. trip and control wiring). This data, together with the modelling knowledge, is sufficient to maintain an accurate representation of the network topology and to draw inference from protection and switchgear activity. Relative likelihood values are given for each type of aspect which a scenario could contain, and may be supplemented with particular values. Finally, some details pertaining to the SCADA system must be provided. Protocol data defines the limits of the fields in the alarm string, and reference details allow particular protection and switchgear equipment to be identified from alarm field contents.

(c) Structure of power system data

In the early stages of development, the network topology was represented by a nodal network, and relays, switchgear and network plant were simply identified by a unique name. While this was satisfactory for the small test systems being used at the time, its limitations arose when applied to a realistic power system.

Power systems were recognized to have an inherent hierarchy: connections to neighbouring power systems relate to the whole system; the local system can be described in terms of substations; each substation contains various pieces of equipment. As a result, network, protection and switchgear equipment is identified by a substation name, and a local name which is unique in that one specific substation. (The substation identifier is also used to describe interconnected substations, e.g. a line could be described by the Prolog term `ss1 to ss2::l1`.)

Secondly, it was observed that, even without focusing down to the level of physical fault position, some protection relays can discriminate between faults within items of plant. For example, transformers often have different earth fault relays protecting their high-voltage (HV) and low-voltage (LV) windings. As a result a lower 'subunit' level was added to the plant data structure. The complete hierarchy for power system plant consists of a substation identifier, a plant item identifier and a subunit name. For example, the HV winding of transformer `t1` in substation `ss1` would be identified by the Prolog term `ss1::t1\\hv_wdg`.

8.3.10 Qualitative simulation

(a) Function

The qualitative simulation knowledge is used by the central diagnosis of the hypothesizer strategy. Its function is to simulate the effect of the genuine occurrence of a fault and failures hypothesis in order to determine the SCADA alarms which would be generated. The simulation can be viewed as consisting of two major parts. The system simulation resolves the primary effect of a system fault, namely the operation of protection relays. The device simulation then simulates the primary activity, and determines and simulates the consequential activity caused by the protection relay's operation. The device simulation knowledge is described with the modelling knowledge in section 8.3.11. The system simulation is described in the following sections.

(b) Reason for being qualitative

The diagnosis could have been developed with a numerical, or quantitative, simulation at its core. The simulation of protection activity would require the current following a fault for various points in the network. This data is not available from the monitoring system. In addition, the computation time to calculate it using fault flows for every hypothesis would be excessive for on-line implementation, while a look-up table of results from off-line analyses would be excessively large. Moreover, a protection engineer would not require a numerical analysis to identify what protection activity would be caused by a fault. A qualitative simulation is therefore used in the identification of suspected system faults and failures. Finer maloperations, like timing problems, could be recognized by having the deeper diagnosis perform a numerical simulation of the central diagnostic conclusion.

(c) The qualitative simulation knowledge

Protection systems are designed with great care given to coordination such that protection nearest the fault location always has the opportunity to operate first. This observation has been used to develop the qualitative simulation. The IKBS always considers protection nearest the hypothesized fault location before more remote devices, and also discriminates between devices at the same location by comparing a priority classification given in each relay's details. The knowledge in the qualitative simulation of fault isolation can best be described by a series of statements. These are listed below.

1. A protection operation is required if a fault path exists, i.e. there exists an electrical connection between the hypothesized fault posi-

tion and a power infeed (machine or supply point) through which fault current can flow.
2. Protection is in a substation within the disturbance area.
3. From the fault location, the plant along the fault path is searched to find CTs.
4. CTs can be on the fault current carrying plant.
5. CTs can be on the HV star point to ground connections of a transformer whose HV winding is connected to a junction which is on the fault path (i.e. an earthing transformer).
6. The highest priority protection relays burdening these CTs have their operation simulated if they can 'see' the fault.
7. Secondary relay and controlled breaker activity is simulated with primary operations.
8. If the same fault path is not isolated then another protection relay is sought – it is a backup operation.
9. A backup operation is identified by looking first at the next highest priority relays and then at the CTs on the plant next along the fault path.
10. When a fault path is isolated, another one is sought.
11. The simulation is complete when no fault path exists if the alarmed device activity has been included in the simulated activity.
12. If the device's activity is not included then an additional failure is required.

After the fault isolation has been simulated, production rules are fired to simulate the post-isolation simulation. This simulated activity may range from simply relay resetting to automatic reclosing activity.

(d) Explanation/understanding facilities

The description of the qualitative simulation illustrates its fundamental nature. The fundamental nature of this knowledge permits the straightforward generation of explanatory details which aid understanding of the situation. A separate specific knowledge base would be required for explanation facilities if a shallower level of diagnosis had been used. An explanation automatically generated by the IKBS (it is not a static body of text) for a particular diagnosis is shown:

```
... Relay rief2 at dumb will not isolate the fault due
to the failure of cb5 at dumb to trip. As a result,
rdef1 at drum operated, rsbef2 at dumb operated and
rsbef1 at dumb operated to isolate the fault.
```

This example describes how the failure of its controlled switchgear caused a protection relay (`rief2` at substation `dumb`) to fail to isolate the fault, and how the relays `rdef1` at `drum`, `rsbef2` at `dumb` and `rsbef1` at `dumb` operated in a backup capacity, resulting in the isolation of the fault.

(c) Power infeed search – finding a fault path

Initially the rule for finding fault paths involved a relatively straightforward depth-first search through the system topology. A fault path was found by searching for plant connected to the hypothesized fault location and then successively looking for plant connected to the new plant until a machine or supply point is found. If the search is stopped (e.g. by an open breaker) then the path is retraced until alternative connected plant is identified, and the search recommenced. Two problems were found with this technique, the first of which was relatively minor.

It was recognized that, having entered a generating station and reached a busbar, the search would often progress out of the station on another circuit before finding the cable to which a machine was connected. Therefore, it was required to bring knowledge of the generator location to the busbar level of the substation topology for early recognition. To resolve this, a rule was created which obtained the route from each local machine to any directly connected busbars. Then, whenever the search for a power infeed enters a station, this rule is invoked and a search for the busbar sections known to be directly connected to machines commenced. In this way the route to power infeed search became more efficient.

The second problem was more serious and was due to the direction in which the power infeed first progresses being merely dependent on the order of topology facts in the database. In some network feeders the power flow is only ever in one direction and protection relays are only installed at the sending end – the protection scheme is effectively open-ended. With a hypothesized fault on such a feeder, the search sometimes failed to identify any protection and sometimes wrongly detected backup protection first.

This problem could occur when a longer path to a power source is found before a path which is shorter. Therefore the use of a breadth-first search appeared to offer a solution: alternative paths would be traced, with progress in each being maintained at approximately the same cumulative impedance. However, in an interconnected network, numerous alternative paths could be created, only to be opened near the fault position by the simulated tripping of one breaker. The depth-first approach is much more efficient in this respect.

Since the problem arises with plant which is only ever used to supply power in one direction, plant of this type could be identified by a fact in the knowledge base. When the hypothesized fault is on such plant, the first fault path search would always be made in the direction given in the fact.

However, a more effective solution was developed and adopted. A load flow is performed once off-line with normal loading conditions, and the results used to identify the subunits from which real current flows into plant items. Then, at all stages of a search, connections to plant in the reverse direction of normal power flow are always attempted first. The result is that, not only is the aforementioned problem resolved, but a power infeed will generally be found more quickly than by the original 'blind' search.

The load-flow program was developed according to the fast decouple algorithm [40] and coded in C. Although the load-flow analysis requires network data in a different form to that held for fault diagnosis, Prolog rules were created to automatically translate the knowledge base network details into the required node and branch data. This translation process creates nodes from busbar and junction sections, three terminal lines, transformers with three power windings and machines, identifies corresponding branch end facts, merges adjacent nodes, and forms the necessary node power and voltage data and branch impedance data. A foreign language interface in the Prolog system is used to initiate the load-flow analysis, but the translated network data and the filtered load-flow results are passed by files. For each branch, the load-flow results file identifies the node at the end from which real current flows. Then, for every plant item along a branch, the subunits on the side of the infeeding node are stored in special facts by Prolog rules. These are the facts which are used on-line.

8.3.11 Modelling knowledge

The modelling knowledge describes knowledge for the various equipment types and relates in particular to system faults, plant internal links, protection detection logic, and protection and switchgear activity.

(a) Faults on plant

Since separate protection relays are often used to protect ground faults and phase faults, the IKBS classifies faults as ground faults or phase faults. Ground and phase faults have different effects on system currents on the supplied side of faulted plant. A phase fault on an item of plant causes an increase in the magnitude of currents in the plant feeding the fault. Ground faults generally cause a lesser increase in current magnitudes, but also introduce a residual component in the currents. The knowledge in the IKBS therefore dictates that:

1. **high phase_effects** are produced by phase faults.
2. **neutral_effects** and **average phase_effects** are produced by ground faults.

In addition to location and generic type, faults are further described by their permanence. Both permanent and temporary faults are hypothesized if the initiating alarmed protection relay belongs to a delayed autoreclose scheme. Otherwise, only a (permanent or temporary) fault is dealt with.

(b) Plant internal links

While the network topology describes the connections between different items of plant, the connectivity between subunits of the same item of plant is generally a characteristic of the plant type. Since various aspects of the diagnosis need to search through the network on different bases, modelling knowledge is included to define the characteristic internal links between transformer windings, between line ends etc. The basis of these internal links is not restricted to physical connectivity.

1. Three-phase connection: this 'connection' is by electrical/physical connectivity or magnetic coupling (i.e. between windings of the same transformer).
2. One-phase connection: this criterion restricts the involved equipment to single phase plant or single phase subunits.
3. **neutral_effects**: a link from one subunit to another exists if a residual component in the current flowing through the first subunit would result in a corresponding effect on the current in the other subunit.
4. **phase_effects**: links are made between subunits if an increased current magnitude in one would result in increased magnitude current flowing through the other subunit.

The derivation of the effect propagation links based on the three-phase electrical connection of subunits is trivial since the currents flowing through the plant remain unchanged. This does not apply to transformers: current propagation is due to magnetic coupling. An increased current magnitude (**phase_effects**) in any transformer winding will induce an increased magnitude field which, in turn, induces an increased current in all windings coupled by this field (i.e. ampere turn balance maintained). Hence, the **phase_effects** propagate between winding subunits for all winding arrangements. They do not, however, propagate to winding neutrals since the net field is zero.

In contrast, the propagation of **neutral_effects** is dependent on the winding arrangement. A grounded star provides a path for a residual component to flow to ground. However, residual components will only flow to ground if another winding maintains ampere turn balance by providing a route for residual currents to flow to ground or circulate. The rule linking a winding and its neutral by `neutral_effects` checks that this ampere-turn balance is maintained by another winding. A rule permits a `neutral_effects` link between different windings if there is a path to ground on both windings.

(c) Protection detection logic

Modelling knowledge is included for each protection type. Facts define the measure by which a protection type detects a fault condition using the terms `phase_effects` or `neutral_effects`, and average or high. Knowledge is also included to determine the area of the network protected by a particular device. The unit protection types are fundamentally defined by the plant enclosed by the associated CTs.

Non-unit protection relays burdening CTs on three-phase plant can detect faults if the measured current, sum of measured currents, or voltage transformer (VT) output to CT output ratio satisfy certain conditions. An accurate model based on extensive numerical computations is not an option due to time constraints. Instead, the IKBS uses two conditions which must be met. The first is that CTs are between the fault location and a power infeed (or on a transformer neutral), and is met in the qualitative simulation knowledge. The second condition is that the fault is located within a protected area of the network. Modelling knowledge defining this area uses cumulative impedances. Basically, plant is progressively identified away from the relaying point, and the impedances of the connected plant added until the cumulative impedance satisfies a condition which is set for the relay type; all such paths form the protected area (ignoring the effect of load current and contributions via branching points in this way limits the accuracy of the modelling knowledge).

(d) Protection and switchgear activity

Modelling knowledge defines how protection and switchgear activity is simulated using the following.

1. Actual activity: this relates to activity which, it is believed, would have taken place under the given hypothesized conditions.
2. Default activity: this is used where the system response to an event is as yet unclear due to there being alternative processes which may occur. For example, in a two-way intertrip scheme, a breaker may trip due to the operation of an intertrip receive relay or before this due to its local protection operating.

3. Possible activity: this is used to cater for the multiple trip/reclose cycles which might occur before an autoreclose scheme successfully extinguishes a temporary fault.

The relay activity model simulates both the protection relay's operation and the directly caused secondary activity. Any device which should operate but is a known or hypothesized failure does not have its operation simulated. The protection relay's trip and control wiring is traced in a depth-first manner and, provided the intervening auxiliary switch logic is satisfied and the controlled device is not a failure, secondary relay and switchgear activity is simulated. When switchgear is reached and simulated or auxiliary switch logic fails or a failed device is encountered, the reasoning traces back along the wiring in search of more secondary activity. In addition, post-isolation production rules have been created in order to simulate the resetting of relays and to simulate auto-reclose activity.

8.4 DEMONSTRATION OF THE DEVELOPED SYSTEM

8.4.1 Case study

Many cases were tested on different portions of network during the development of the IKBS described in the previous section. Both hypothetical and genuine networks were used. A utility protection engineer helped in the creation of some of these studies. Figure 8.1 shows a portion of network containing 132 kV, 33 kV and 11 kV plant.

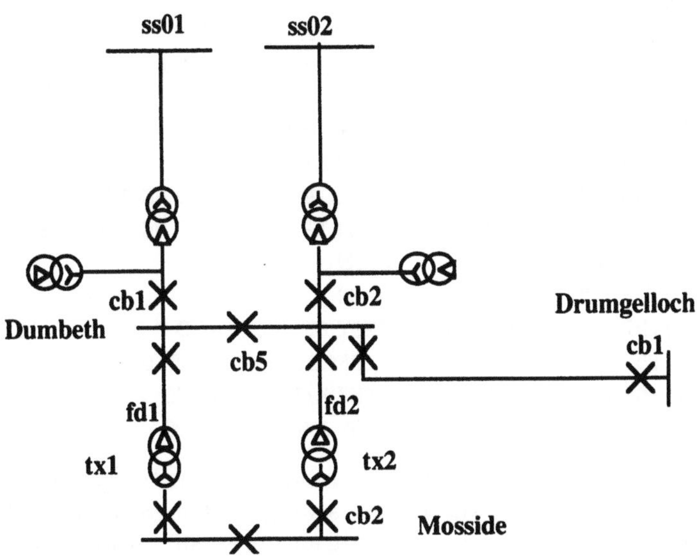

Figure 8.1 A portion of network.

Substations ss01, ss02 and Drumgelloch are defined as supply points. Both 33 kV transformer feeders between substations Dumbeth (dumb) and Mosside (moss) are protected against ground faults by instantaneous earth fault relays at substation Dumbeth which have intertripping to the remote end breakers. Standby earth fault relays measure the current in the neutrals of earthing transformers connected to the infeed circuits of substation Dumbeth. There is a directional earth fault relay at substation Drumgelloch (drum). Clearly, this is not a description of the complete protection system, but rather is some details relevant to this case study.

This case study relates to the occurrence of a ground fault on the second transformer feeder of substation Dumbeth for the case where breaker cb5 at dumb remains stuck closed. Under this circumstance, the instantaneous earth fault relay operates, its trip relay operates and attempts to trip cb5, an intertrip is sent to the remote end which results in the tripping of the breaker at Mosside. The fault persists because breaker cb5 at dumb fails to trip. As a result, the directional earth fault relay at Drumgelloch and the standby earth fault relays at Dumbeth operate, and the fault is isolated when their controlled breakers trip. If the SCADA system functioned perfectly, the following alarms would be received:

```
10:16:00:32 DUMB  MOSS NO2  INST EF PROT OPTD              ON
10:16:00:33 DUMB  MOSS NO2  TRIP RELAYS TO BE RESET-E       ON
10:16:00:39 MOSS  T2        INTERTRIP RECEIVE OPTD          ON
10:16:00:42 MOSS  CB2       OPEN
10:16:00:84 DRUM  DUMB      DIRECTIONAL EF OPTD             ON
10:16:00:85 DRUM  DUMB      TRIP RELAYS TO BE RESET-E       ON
10:16:00:88 DRUM  CB1       OPEN
10:16:04:44 DUMB  SS02      STBY E/F PROT OPTD              ON
10:16:04:44 DUMB  SS01      STBY E/F PROT OPTD              ON
10:16:04:45 DUMB  SS02      T2 TRIP RELAYS TO BE RESET-E    ON
10:16:04:45 DUMB  SS01      T1 TRIP RELAYS TO BE RESET-E    ON
10:16:04:48 DUMB  CB2       OPEN
10:16:04:48 DUMB  CB1       OPEN
10:16:04:52 DUMB  MOSS NO2  INST EF PROT OPTD              OFF
10:16:04:52 DRUM  DUMB      DIRECTIONAL EF OPTD             OFF
10:16:04:53 DUMB  SS02      STBY E/F PROT OPTD              OFF
10:16:04:53 DUMB  SS01      STBY E/F PROT OPTD              OFF
```

The above alarms include main operations, remote backup operations, local backup operations, and relay resetting. Although the case studied would not occur frequently, it is not an unrealistic incident, and will demonstrate the type of help which would be made available to an operator.

8.4.2 Hypothesizer diagnosis

No disturbances existed before the above alarms. When the alarm indicating the operation of the instantaneous earth fault relay is received, the IKBS begins generating no-failure hypotheses based on the relay's protection modelling knowledge. The winding arrangement of the transformer at the Mosside end of the feeder is found to 'block' neutral effects with the result that the relay is found to only detect a ground fault on the feeder or on the transformer primary winding. These two single fault hypotheses and a false operation hypothesis are successfully simulated. Since they describe predicted secondary activity, the next three alarms strengthen all three scenarios.

The 'gut reaction' then identifies the directional earth fault relay alarm as unexpected but potentially related. Only breaker `cb5` at `dumb` is a possible failure since it is the only device with predicted activity (disregarding resetting and reclosing alarms). The two existing single fault hypotheses therefore have the breaker failure added to form new hypotheses which are then simulated. Successful partial scenarios are formed from the simulation results. The partial scenarios are each copied as complete scenarios since every disturbance alarm is already accounted for. A false-alarm partial scenario is then generated and combined with the old false-alarm scenario to form an updated false-alarm scenario and with the two single-fault scenarios to form two 'fault-plus-false-alarm' scenarios. The remaining alarms strengthen the two single-fault-single-failure scenarios, whereas the two remaining primary alarms are included as false activity in the other scenarios. After all the listed alarms have been received, the resulting ranking table appears as:

```
Last alarm time: 10:16:04.53

5 possibilities

1. There was a ground fault_on dumb to moss::fd2 which_is
(permanent or temporary),
which the protection equipment cleared.
However cb5 at dumb failed to trip.

2. There was a ground fault_on moss::tx2\\hv_wdg which_is
(permanent or temporary),
which the protection equipment cleared.
However cb5 at dumb failed to trip.

3. There was a ground fault_on dumb to moss::fd2 which_is
(permanent or temporary),
which the protection equipment is attempting to clear.
False activity: rsbef1 at dumb reset, rsbef2 at dumb
reset, rdef1 at drum reset,
```

```
rsbef1 at dumb operated, rsbef2 at dumb operated and
rdef1 at drum operated.
```

Clearly, the developed hypothesizer gives the correct result. When displayed in detailed form, the top scenario also conveys the explanation details contained in the scenario:

```
1. There was a ground fault_on dumb to moss::fd2 which_is
(permanent or temporary),
which the protection equipment cleared.
Supporting alarms: t5 at dumb operated, rief2 at dumb
operated, cb2 at moss tripped,
tirrt2 at moss operated, cb1 at drum tripped, t1 at drum
operated, rdef1 at drum
operated, cb1 at dumb tripped, t1 at dumb operated,
rsbef1 at dumb operated, cb2 at
dumb tripped, t2 at dumb operated, rsbef2 at dumb oper-
ated, rief2 at dumb reset, rdef1
at drum reset, rsbef1 at dumb reset and rsbef2 at dumb
reset.
However cb5 at dumb failed.
Relay rief2 at dumb will not isolate the fault due to
the failure of cb5 at dumb to
trip. As a result, rdef1 at drum operated, rsbef2 at
dumb operated, rsbef1 at dumb
operated to isolate the fault.
```

8.5 FINAL REMARKS

Some remarks can be made based on the experience of building the described IKBS. Firstly, the feasibility of a fault diagnostic IKBS has been proved by the work of the authors and of other researchers. The worth of systems that can cope with real-world problems such as simultaneous disturbances and failed equipment is now being widely accepted.

In contrast with many other IKBSs, a significant level of network knowledge is included in the developed IKBS – especially that using the hypothesizing strategy. In this sense it may be described as being 'model-based'. The developed IKBS has a significant level of understanding due to this knowledge. Moreover, the knowledge is fundamental, essential or generic in nature, with the result that the IKBS is very portable. In addition, the level of understanding has provided the means for a limited explanation of fault-and-failure diagnoses without the addition of dedicated explanatory rules. Furthermore, the level of understanding included means that more knowledge can be easily added in order to progress the reasoning further, e.g. into restorative advice.

The importance of adequately structuring the knowledge has been recognized. This was recognized during the research when the problems of the scale of genuine power systems were considered. A hierarchical structure has been used for the diagnostic knowledge, diagnostic conclusions and the specific network details. This structuring has made the diagnosis more focused. The worth of the logic rule-based paradigm, and also of Prolog as a development tool has been proven. The knowledge is transparent since the language is symbolic and can be recognizably close to natural language. An evolutionary development cycle can therefore be used, and this has been employed. The feasibility of Prolog as an implementation language has yet to be fully evaluated. However, the developments in distributed control centre computing facilities have independently provided an environment suitable for the implementation of IKBSs. What remains to be established is whether or not Prolog can provide the speed of performance required for a real-time application.

The management of uncertainty in a fault diagnostic IKBS is problematic. Techniques range from counting maloperations to mathematically based theories. In the light of the limited data available and the nature of the domain, 'relative likelihood' has been developed as a viable alternative.

It has been found that the building of the databases necessary for a demonstration of a working IKBS takes a significant part of the development time. The benefits to be had from maintaining a complete protection database, a symbolic connectivity database, plant history database etc. are only truly realized when used with or integrated with an IKBS such as that developed. The development of an IKBS will therefore, of necessity, involve at least the partial creation of system databases – this is a time-consuming task.

The prototyping method of development does hold benefits over a more traditional design–coding–debugging approach as was claimed by Wollenberg and Sakaguchi [8]. However, it must be used with care. Although tested on small test systems, the development must not progress without bearing the ultimate implementation in mind. For example, data structures were influenced by the problems of the scale of genuine power systems. If all attention had been given to the prototype system, problems with unsuitable data structures may have appeared later in the development.

Use is made of a load-flow routine to supplement the knowledge base of the developed IKBS. There is significant potential for algorithmic analysis routines to be used co-operatively with knowledge based systems. Two observations may be made regarding this aspect of the work. Firstly, the data required by knowledge- and algorithm-based systems may relate to the same application system but be of different types – automatic translation routines were developed to minimize duplicated effort and ease maintenance. The second observation is that seldom is the vast amounts of data generated by the likes of load-flow routines of interest. Since only certain features in that data were of importance, placing the filtering routines on the algorithm side of the interface dramatically reduced the volume of transferred data.

It has been found that literal messages are particularly powerful. Although they are longer than corresponding messages which merely express the information they contain, the literal messages output by the fault diagnosis have been well received and the information quickly assimilated by engineers. The main limitation is in expressing the location of a disturbance or maloperated device; a role for which graphics are particularly suited. It is believed that a graphical representation coupled with literal explanation would be the most user friendly and effective means of informing power system operators of diagnostic conclusions.

REFERENCES

1. A Working Group of the Relaying Practices Subcommittee of the IEEE Power System Relaying Committee. (1989) Summary of the special publication 'Applications of fault and disturbance recording devices for protective system analysis'. *IEEE Transactions on Power Delivery*, **PWRD-4**(3), 1625–30.
2. Dy Liacco, T.E. and Kraynak, T.J. (1969) Processing by logic programming of circuit-breaker and protective-relaying information. *IEEE Transactions on Power Apparatus and Systems*, **PAS-88**(2), 171–75.
3. Prince, W.R., Wollenberg, B.F. and Bertagnolli, D.B. (1989) Survey on excessive alarms. *IEEE Transactions on Power Systems*, **PWRS-4**(3), 950–56.
4. Smith, H.L. and Modzelewski, T.J. (1989) Enhancing energy management systems with advanced RTU capabilities. *IEEE Computer Applications in Power*, October, 26–29.
5. Weedy, B.M. (1984) *Electric Power Systems*, Wiley.
6. Fukui, C. and Kawakami, J. (1986) An expert system for fault section estimation using information from protective relays and circuit breakers. *IEEE Transactions on Power Delivery*, **PWRD-1**(4), 83–90.
7. Claus, J., de Viron Dongier, F., Baudrux, M., Lienart, P., Van Miegroet, P. and Maun, J.C. (1991) Processing the indications of operation and alarm in the electrical networks, in *Colloquium of CIGRE Study Committee 34 – CIGRE24, Preferential Subject on Development and Applications of Expert Systems to Protection and Local Control Systems*, Brazil, September.
8. Wollenberg, B.F. and Sakaguchi, T. (1987) Artificial intelligence in power system operations. *Proceedings of the IEEE*, **75**(12), 1678–85.
9. Girgis, A.A. and Johns, M.B. (1989) A hybrid expert system for faulted section identification, fault type classification and selection of fault location algorithms. *IEEE Transactions on Power Delivery*, **PWRD-4**(2), 978–85.
10. Watanabe, K. and Oobe, N. (1991) On-line data analysis system with remote data transmission, in *Colloquium of CIGRE Study Committee 34 – CIGRE24, Preferential Subject on Development and Applications of Expert Systems to Protection and Local Control Systems*, Brazil, September.

11. Kirschen, D.S., Wollenberg, B.F., Irisarri, G.D., Bann, J.J. and Miller, B.N. (1989) Controlling power systems during emergencies: the role of expert systems. *IEEE Computer Applications in Power*, April, 41–45.
12. Hotta, K., Nomura, H., Takemoto, H., Suzuki, K., Nakamura, S. and Fukui, S. (1990) Implementation of a real-time expert system for a restoration guide in a dispatching centre. *IEEE Transactions on Power Systems*, **PWRS-5**(3), 1032–38.
13. Cardozo, E. and Talukdar, S.N. (1988) A distributed expert system for fault diagnosis. *IEEE Transactions on Power Systems*, **PWRS-3**(2), 641–46.
14. Talukdar, S.N., Cardozo, E. and Leao, L. (1986) TOAST: the power system operators assistant. *IEEE Computer*, **19**(7).
15. Talukdar, S.N., Cardozo, E. and Perry, T. (1986) The operators assistant – An intelligent, expandable program for power system trouble analysis. *IEEE Transactions on Power Systems*, **PWRS-1**(3), 182–87.
16. Bramer, M.A., Pearce, J.S., Muirden, D., Platts, J.C. and Vipond, D.L. (1988) FAUST – an expert system for diagnosing faults in an electricity supply system, in *Research and Development in Expert Systems 5* (eds B. Kelly and A. Rector), Cambridge University Press.
17. Cockburn, D., McDonald, J.R., Burt, G., Brailsford, J.R., Beaton, J. and Lo, K.L. (1991) Expert systems for on-line fault diagnosis in electrical power networks, in *11th International Conference on Electricity Distribution (CIRED)*, Liege, April.
18. Esp, D.G. and Cahill, E.T. (1991) A real time expert system for fault diagnosis on a transmission network, in *3rd International Conference on Power System Monitoring and Control*, London, June pp. 237–40.
19. Jeyasurya, B., Venkata, S.S., Vadari, S.V. and Postforoosh, J. (1990) Fault diagnosis using substation computer. *IEEE Transactions on Power Delivery*, **PWRD-5**(2), 1195–1201.
20. Ares, J.-M. and Girouard, P. (1990) GESTAL: a specialised tool to build real-time alarm processing and fault diagnosis expert systems for power network control centres, in *Expert Systems Applications for the Electric Power Industry* (ed. J.A. Naser), Vol. 1 of 2, Hemisphere Publishing, pp. 679–89.
21. Balakrishnan, R. and Pahwa, A. (1990) A computer assisted intelligent storm outage evaluator for power distribution systems. *IEEE Transactions on Power Delivery*, **PWRD-5**(3), 1591–97.
22. Hsu, Y.-Y., Lu, F.C., Chien, Y., Liu, J.P., Lin, J.T., Yu, H.S. and Kuo, R.T. (1991) An expert system for locating distribution system faults. *IEEE Transactions on Power Delivery*, **PWRD-6**(1), 366–72.
23. Okamoto, H., Yokoyama, A. and Sekine, Y. (1990) A real-time expert system for fault section estimation using cause-effect network, in *Proceedings of the 10th Power Systems Computation Conference*, Graz, Austria, August, pp. 905–12.
24. Partanen, J., Koppanen, T. and Jarventausta, P. (1990) An expert system for fault diagnosis of medium-voltage distribution networks, in *Proceedings of the 10th Power Systems Computation Conference*, Graz, Austria, August, pp. 1080–87.
25. Marathe, H.Y., Liu, C.-C., Tsai, M.-S., Rogers, R.G. and Maurer, J.M. (1991) An on-line operational expert system with data validation capabilities. *IEEE Transactions on Power Systems*, **PWRS-6**(2), 882–89.
26. Tomsovic, K., Liu, C.-C., Ackerman, P. and Pope, S. (1987) An expert system as a dispatchers aid for the isolation of line section faults. *IEEE Transactions on Power Delivery*, **PWRD-2**(3), 736–43.
27. Fauquembergue, P. and Esquevin J.V. (1991) An expert system as an on-line

aiding tool for the maintenance of protection systems, in *Colloquium of CIGRE Study Committee 34 – CIGRE24, Preferential Subject on Development and Applications of Expert Systems to Protection and Local Control Systems*, Brazil, September.
28. Amantegui, J., Corera, J., Echavarri, J., Barandiaran, J., Laresgoiti, I. and Perez, J. (1991) Expert system for fault analysis in electrical networks. Operating experience and further developments, in *Colloquium of CIGRE Study Committee 34 – CIGRE24, Preferential Subject on Development and Applications of Expert Systems to Protection and Local Control Systems*, Brazil, September.
29. Keronen, J.J. (1989) An expert system prototype for event diagnosis and real-time operation planning in power system control. *IEEE Transactions on Power Systems*, **PWRS-4**(2), 544–50.
30. Keronen, J.J. (1990) EKA: an expert system for real-time operation planning and event analysis in electric power networks, in *Expert Systems Applications for the Electric Power Industry* (ed. J.A. Naser), Vol. 1 of 2, Hemisphere Publishing, pp. 447–63.
31. Kaminaris, S.D., Machias, A.V. and Papadias, B.C. (1991) An intelligent tool for distribution substations troubleshooting and maintenance scheduling. *IEEE Transactions on Power Delivery*, **PWRD-6**(3), 1038–44.
32. Protopapas, C.A., Psaltiras, K.P. and Machias. A.V. (1991) An expert system for substation fault diagnosis and alarm processing. *IEEE Transactions on Power Delivery*, **PWRD-6**(2), 648–55.
33. McDonald, J.R., Burt, G.M. and Young, D.J. (1992) Alarm processing and fault diagnosis using knowledge based systems for transmission and distribution network control. *IEEE Transactions on Power Systems*, **PWRS-7**(3), 1292–98.
34. Liu, C.-C., Sforna, M. and Miao, H. (1996) On-line fault diagnosis using sequence-of-events recorder information, in *Proceedings of the International Conference on Intelligent System Applications to Power Systems (ISAP96)*, pp. 339–44.
35. Vazquez M.E. *et al.* (1996) On-line knowledge based system for fault section diagnosis in control centers, in *Proceedings of the International Conference on Intelligent System Applications to Power Systems (ISAP96)*, pp. 232–36.
36. Tang, S.K., Dillon, Th.S. and Khosla, R. (1996) Application of an integrated fuzzy, knowledge based connectionist architecture for fault diagnosis in power systems, in *Proceedings of the International Conference on Intelligent System Applications to Power Systems (ISAP96)*, pp. 188–93.
37. Quintus Computer Systems. *Quintus Prolog Users Guide and Reference Manual.*
38. Ng, K.-C. and Abramson, B. (1990) Uncertainty management in expert systems. *IEEE Expert*, April, pp. 29–47.
39. Chang, C.S. *et al.* (1996) Power system fault diagnosis using fuzzy sets for uncertainties processing, in *Proceedings of the International Conference on Intelligent System Applications to Power Systems (ISAP96)*, pp. 333–38.
40. Arrillaga, J., Arnold, C.P. and Harker, B.J. (1983) *Computer Modelling of Electrical Power Systems*, Wiley.

9

Other applications in power system operation and control

Jerzy S. Zielinski

9.1 VOLTAGE/REACTIVE POWER CONTROL

In addition to the problems considered in Chapters 4 to 8 of this book, voltage/reactive power (VAR) control and emergency control and restoration need special attention in power system operation and control. The above tasks need to be solved on the occurrence of power system outages, but the voltage/VAR problem also arises under steady-state conditions. For this reason, separate intelligent knowledge based systems (IKBSs) have been developed for VAR and emergency control.

In this chapter's references, a number of IKBSs in different development stages have been listed, some of which will be presented below.

Barruncho *et al.* [1] report the implementation and operational experience of an IKBS for VAR control in a real-time environment. This reactive management and voltage monitoring and control (RVMC) IKBS has been developed in collaboration between the Technical University of Lisbon and the National Power System Utility (EDP) in Portugal [2], and is embedded in an existing energy management system. A human–machine interface and a number of specialized tools support the RVMC in its performance of the following synchronized tasks:

1. database maintenance after each valid state estimated solution;
2. running the reactive optimal power flow (ROPF) process for the global redefinition of the desired voltage profile and reactive management-related variables;
3. load bus voltage sensitivity processing, executed under close co-operation with a voltage control expert system (VCES);
4. running the VCES after each valid state estimation.

The VCES uses a rule-based knowledge representation with the following rule goal classes: monitoring, controlling, prediction and validation. Some benefits have been reported from the application of the system, e.g. the ROPF function has achieved an average reduction of accumulative active losses of about 5–7% (of 2400 MW h).

A very interesting report [3] is based on a questionnaire from 31 respondents in 14 countries. Some of the most interesting data from these responses is listed as follows.

1. Systems in development: planning, three IKBSs; prototype, five IKBSs; implemented in real time, one IKBS; commercial product, one IKBS.
2. Control algorithms used: forward chaining; sensitivity analysis; heuristic search; broadcasting algorithm and bidding concept; forward and backward chaining; load flow; contingency analysis and sensitivity analysis controlled by IKBS rules.

Another paper of interest [4] considers European IKBS applications in power systems. Following some short remarks on the role of computers in power systems and the motivation for developing knowledge based systems, the authors describe major research projects in Europe, including among these one voltage and VAR control IKBS [1] undergoing field tests, two prototypes and one in the feasibility study stage.

The IKBS for voltage collapse detection and prevention (KBVCDP) [5] consists of power-flow divergence monitoring and indicator selection modules and has been tested on the IEEE 57 bus power system.

Some authors have developed new algorithms for solving VAR problems. Liu *et al.* [6] have applied modified inductive learning, in which classification rules are represented by a decision tree, to solve the problem. Test cases were performed using the IEEE 30 bus system with modified load characteristics. Wehenkel and Pavella [7] used the decision tree approach to solve the power system security assessment and voltage security problems. Tomsovic and Barruncho [8] used a Petri-net model of a rule-based IKBS to identify properties of the rule base. Haida and Akimoto [9] applied genetic algorithms (probabilistic search procedures for optimization) to the power system voltage control problem. Two experiments on a model system demonstrated the effectiveness of the approach for more difficult optimization problems.

9.2 EMERGENCY CONTROL AND RESTORATION

Emergency control in contemporary bulk power systems relates in general to the restoration procedure. There are different subgoals in the restoration problem.

1. In a bulk power system it is necessary to restore the load flow in trunk transmission lines.
2. In a distribution and/or subtransmission system loss reduction through feeder reconfiguration is of great significance.

These subgoals divide all papers concerned with IKBS applications to restoration into three groups: the first considers bulk power system restoration, the second relates to subtransmission system restoration and the third to distribution systems. The necessity of using IKBSs as the new tool for aiding black start and restoration situations is stressed by Hayashi *et al.* [10].

Matsumoto *et al.* [11] present a bibliographical review of the application of knowledge based systems to bulk power system, subtransmission system and distribution system restoration. The power system restoration solution may use one the following techniques: operational aids methodology; restoration level methodology; system development methodology. According to the operational aids methodology, guidance systems or decision support systems are needed. The restoration level methodology consists of state recognition, restoration plan, and restorative operation stages. In the system development methodology the knowledge acquisition and integration with an existing environment are important. Three approaches can be considered for integration with an existing environment: appended configuration, embedded configuration and back end configuration.

Shimakura *et al.* [12] have developed a prototype IKBS for making restoration plans for a bulk power system. The knowledge base consists of about 200 OPS83 rules, and includes general purpose restoration knowledge which does not depend on pre-outage system states. Test results have been confirmed by veteran operators, and have taken between 6 and 100 s processing time on a 7 MIPS engineering workstation (depending on load).

A detailed description of the twice-field-tested IKBS developed by Kuroda *et al.* is given in reference 13 (this system was earlier reported in reference 14). This IKBS provides security monitoring, restoration guidance and a human–machine interface. The restoration guidance system consists of 500 rules and about 200 functions written in OPS83 and C. The IKBS has been tested with a 77/33 kV radial network containing 120 power stations/substations, major extra-high-voltage loads and about 80 transmission routes. Processing time depends on the realized functions as follows: restoration guidance control, 0.1 s; faulty equipment identification, 5 s; restoration planning, 19 s; restorative operation guidance, 3 s.

Liu *et al.* [15] describe the following knowledge based restoration strategies for a distribution system: group planning, group modification, multilayer load transfer, unrestorable zone modification and switching operation ordering. A forward-chaining inference mechanism is used with frames and production rules in the prototype IKBS, which has been implemented on a PC.

Additional papers present general considerations of IKBS developments, relating in the main to prototype IKBSs for power system restoration. For example:

1. prototype rule-based IKBS for the initial stage of restoring an EHV power system from system blackout [16,17];
2. prototype rule-based IKBS for distribution network restoration [18–21];
3. prototype rule-based IKBS for a 77 kV subtransmission system [22];
4. prototype IKBS using fuzzy reasoning when there is more than one candidate operation for restoration [12].

9.3 REAL-TIME PROCESSING

Previous considerations (Chapter 3) have shown that real-time processing is a very important problem in power system operation. In a search of the literature generated over the last few years one can find a number of papers considering the real-time tasks in power system. Most of them deal with the following problems:

1. real-time power system simulation for testing relays [23–25], for electromagnetic transients [26,27], for general-purpose problem solving [28,29] and for teaching [30];
2. the development of shells or tools necessary for real-time problem solving [31–33];
3. the application of real-time SCADA data to network switching planning [5], load frequency control [34], load shedding [35], VAR control [36,37], substation fault diagnosis [38] and general-problem solving [39–42].

Those papers in which the real-time processing is linked with an IKBS or other AI approach relate to the following problems:

1. power system monitoring [43,44];
2. nuclear power plant monitoring of all the electrical power supplies [45], eddy current inspection of steam generator tubes [46] and plant diagnosis [47];
3. process monitoring in a coal-fired power plant [48] and turbine diagnosis in a thermal power plant [49];
4. fault diagnosis [50–52];
5. feeder operating conditions [53] and restoration [54];
6. voltage and reactive power control [55].

A review of the papers devoted to real-time problem solving using AI tools indicates the following important conclusions for practitioners.

1. Most of the market tools are inappropriate for real-time applications, necessitating the development of special tools or hybrid IKBSs (i.e. knowledge based systems and ANN).
2. Most of the systems are in the early stages of development, with only a very few in practical operation (see also section 3.3).
3. The development process of a real-time AI system is a time-consuming process and needs special attention.

9.4 THE TESTING OF EQUIPMENT

In addition to the areas of IKBS application listed in section 2.2, IKBSs are destined to be used in equipment testing. Clearly the reliability of power system operation depends on the reliability of the system's components, and this necessitates the proper testing of protection relays, generators, transformers, lines, circuit breakers etc. In Part One the Further Reading list contains a number of IKBS applications in the testing domain, and in this chapter some examples concerning the testing of equipment important to power system operation will be considered.

9.4.1 Transformers

Transformers, and especially large power transformers, are essential and expensive elements of a power system, and require careful design, construction and operation. During operation, the diagnosis of a transformer's state is of great importance. This assessment is based on a number of measured quantities, but also requires a number of other parameters to be taken into account, such as transformer age, cooling system, operating conditions. Therefore, a correct diagnosis is a result of a knowledge based process as opposed to a simple assessment of measured values. For these reasons an IKBS approach to transformer diagnosis is highly suitable.

A number of papers describe knowledge based systems for transformer diagnosis: an IKBS concept [56] and IKBS prototypes [57,58]. The adaptive monitoring system of reference 58 is the most interesting among these, and is now described.

The Laboratory for Electromagnetic and Electronic Systems at MIT has undertaken a research programme in transformer monitoring under electric utility sponsorship which incorporates: sensors and sensor signals; modelling of the operation of transformer subsystems; monitoring system software and hardware; the testing of the sensors and systems on a 50 kV A transformer. The adaptive monitoring system consists of sensors, and modules with primary and secondary control buffers, and provides a trend analysis as its output. Each module contains the following blocks: data conversion and validation; adaptive modelling; error computation and anomaly detection. A loosely-coupled multitasking, multiprocessing computer system supports the adaptive monitoring system. Laboratory implementation of the system shows that it is much more sensitive than standard threshold level detection techniques.

9.4.2 Protection relays

The evaluation of relay test results is a very important but time-consuming task, and one for which the application of an IKBS could be very helpful. Sun *et al.* [59] present RETEX as an IKBS for the analysis of relay testing data. RETEX is a prototype which is limited to impedance relays. The development tool used was NEXPERT Object, which can interface with programs written in C, Fortran and other conventional languages. RETEX will be further developed to test other types of relays (travelling wave, overcurrent, bus, differential and transformer differential).

Protection scheme design is another process for which a knowledge based approach could be used. Hayes [60] describes an AI approach using the constraint-directed search method for differential relay setting design and Lee and Liu [61] discuss IKBS techniques for the investigation of protection coordination.

9.4.3 Substations

The application of IKBSs in substation operation within the plant testing domain has also been considered. For example, an IKBS prototype [62], implemented on a portable PC using a LISP-based shell, is used in the acoustic diagnosis of gas-insulated switchgear (GIS). Another important consideration is the complete decentralization of IKBS data processing, testing and monitoring to the substation [63,64].

The IKBSs applied in the testing of power system equipment which have been presented above are only part of a long list of IKBSs which will be developed in the near future.

9.5 MANAGEMENT SYSTEMS

Consideration has been given in Chapters 3 to 8 to IKBS applications in power system design, test and operation. All of these IKBS functions relate in some way to energy management systems, which contain both technological and business functions. It is important to remember that IKBSs are applied in all fields of business (see the Further Reading list of Part One and also Waterman's book [65]).

9.6 ELECTRICAL AND OTHER INDUSTRIES

Industries other than the electrical supply industry, especially those which supply equipment to utilities, are interested in IKBS applications in management, manufacturing, testing and diagnosis. Although this book has concentrated on power system applications of IKBSs, many of the considerations, problems and benefits described herein are equally applicable to IKBS applications within these other industries.

REFERENCES

1. Barruncho, L., Sucena Paiva, J.P., Pestana, R., Vidigal, A. and Liu, Ch.-Ch. (1991) Reactive management and voltage monitoring and control: on-line experience at the Electricidade de Portugal Control Centre, in *Third Symposium on Expert Systems Application to Power Systems*, Tokyo, April pp. 165–72.
2. Barruncho, L., Sucena Paiva, J.P. and Liu, Ch.-Ch. (1989) Voltage/VAR control optimization and knowledge oriented approach: an application for the present run time environments, in *Second Symposium on Expert System Application to Power Systems*, Seattle, USA, July, pp. 451–57.
3. Expert system applied to voltage and VAR control. Final report. *Electra* No. 139, December 1991, 109–31.
4. Germond, A.J. and Niebur, D. (1992) Survey of knowledge based systems in power systems: Europe. *Proceedings of the IEEE*, **80**(5), 732–44.
5. Keronen, J. and Lehtonen, M. (1986) A knowledge based system for real time operators assistance in control of power transmission network, in *Symposium on Expert Systems Application to Power Systems*, Stockholm–Helsinki, August, pp. 16-1–16-4.
6. Liu, Ch.-Ch., Wang, S.-M., Wong, L.W., Marathe, H.Y. and Lauby, M.G. (1989) A self-learning expert system for voltage control of power systems, in *Second Symposium on Expert System Application to Power Systems*, Seattle USA, July, pp. 462–68.
7. Wehenkel, L. and Pavella, M. (1993) Decision tree approach to power system security assessment. *International Journal of Electrical Power and Energy Systems*, **15**(1), 13–36.
8. Tomsovic, K. and Barruncho, L. (1989) Petri net representation as an evaluation method for a voltage control expert system, in *Second Symposium on Expert System Application to Power Systems*, Seattle USA, July, pp. 409–15.
9. Haida, T. and Akimoto, Y. (1991) Voltage optimization by using genetic algorithms, in *Third Symposium on Expert Systems Application to Power Systems*, Tokyo, April, 375–80.
10. Hayashi, T. *et al.* (1993) Modelling and simulation of black start and restoration of electric power systems. *Electra* No. 147, April, 21–42.
11. Matsumoto, K., Sakaguchi, T., Kafka, R.J. and Adibi M.M. (1992) knowledge based systems as operational aids in power system restoration. *Proceedings of the IEEE*, **80**(5), 689–97.
12. Shimakura, K., Inagaki, J., Matsunoki, Y., Ito, M. and Hori, S. (1992) A knowledge based method for making restoration plan of bulk power system. *IEEE Transactions on Power Systems*, **7**(2), 914–20.
13. Kuroda, K., Takegoshi, Y., Yamashita, K., Ito, M. and Fukui, S. (1991) Verification of a knowledge based restoration guidance system in a dispatching

centre, in *Third Symposium on Expert Systems Application to Power Systems*, Tokyo, April, pp. 256–61.

14. Hotta, K., Nomura, H., Takemoto, H., Suzuki, K., Nakamura, S. and Fukui, S. (1990) Implementation of real-time expert system for a restoration guide in a dispatching centre. *IEEE Transactions on Power Systems*, 5(3), 1032–38.
15. Liu, C.C., Lee, S.J. and Venkata, S.S. (1987) An expert system operational aid for restoration and loss reduction of distribution systems, in *PICA '87*, May, pp. 79–85.
16. Fandino, J., Darnault, P.H., Bigeon, J., Sabonnadiere, J.C., Harmand, Y., Heilbronn, B. and Mondon, E. (1991) An expert system as help for power system restoration after blackout, in *Third Symposium on Expert Systems Application to Power Systems*, Tokyo, April, pp. 487–94.
17. Kakimoto, N., Lin, B. and Hayashi, M. (1991) Restoration of power system from complete outage by expert system, in *Third Symposium on Expert Systems Application to Power Systems*, Tokyo, April, 262–69.
18. Fudo, H., Egawa, S., Sanga, K., Inoue, H. and Imamura, Y. (1991) An expert system for restoration of distribution network, in *Third Symposium on Expert Systems Application to Power Systems*, Tokyo, April, pp. 695–700.
19. Schwarzjirg, H. and Brugger, H. (1991) Network restoration/reconfiguration and load transfer for a distribution network using a knowledge based system, in *Third Symposium on Expert Systems Application to Power Systems*, Tokyo, April, pp. 708–15.
20. Elders, I.M., Burt, G.M., McDonald, J.R. *et al.* (1996) The application of an intelligent system to power network operation and control, in *Proceeedings of the International Conference on Intelligent System Applications to Power Systems (ISAP'96)*, pp. 170–74.
21. Kadar, P. and Mergl, A.K. (1996) CORES – the continuous restoration expert system, in *Proceedings of the International Conference on Intelligent System Applications to Power Systems (ISAP'96)*, pp. 390–93.
22. Fujita, H., Shibazaki, M., Ueki, T., Niimura, T., Takenaka, M. and Matsui, T. (1991) An expert system for subtransmission system restoration, in *Third Symposium on Expert Systems Application to Power Systems*, Tokyo, April, pp. 277–83.
23. McLaren, P.G., Kuffel, R., Wiercks, R., Giesbrecht, J. and Arendt, L. (1991) A real time power system simulator for testing relays, in *Proceedings of the 1991 IEEE Power Engineering Society, Transmission and Distribution Conference*, Dallas, Texas, 22–27 September, pp. 370–75.
24. McLaren, P.G., Kuffel, R., Wierckx, R., Giesbrecht, J. and Arendt, L. (1992) A real-time digital simulator for testing relays. *IEEE Transactions on Power Delivery*, **7**(1), 207–13.
25. Nimmersjö, G., Hillström, B., Werner-Erichsen, O. and Rockefeller, G.D. (1988) A digitally-controlled, real-time, analog power system simulator for closed-loop protective relaying testing. *IEEE Transactions on Power Delivery*, **3**(1), 138–52.
26. Rumpel, D., Venkata, S.S., Pandji, K., Liu, C.C. and Nagdy, N.M. (1990) Real-time database for power systems using language oriented data structure. *IEEE Transactions on Power Systems*, **5**(3), pp. 993–1000.
27. Wang, X. and Mathur, R.M. (1989) Real-time digital simulator of the electromagnetic transients of transmission lines with frequency dependence. *IEEE Transactions on Power Delivery*, **4**(4), 2249–55.
28. Bose, A. and Clemens, K.A. (1987) Real-time modeling of power networks. *Proceedings of the IEEE*, **75**(12), 1607–22.
29. Kokai, Y., Matori, I. and Kawakami, J. (1988) Multiprocessor based generator

module for a real-time power system simulator. *IEEE Transactions on Power Systems*, **3**(4), 1633–39.

30. De La Ree, J. and Latorre, J.A. (1990) PC based real-time simulation for teaching power generation concepts. *IEEE Transactions on Power Systems*, **5**(1), 326–30.
31. Cardozo, E. (1989) Expert systems for real time applications: some design and implementation issue, in *Second Symposium on Expert System Application to Power Systems*, Seattle USA, July, pp. 469–473.
32. Schwab, B.C., Hemmaplardh, K., Sackett, S.A., Lawrence, G.M., Davis, R.N. and Daine, D.N. (1988) Software features of a DBMS kernel in large real-time EMS systems. *IEEE Transactions on Power Systems*, **3**(1), 67–72.
33. Talukdar, S., Ramesh, V.C. and Nixon, J.C. (1991) A distributed system of control specialists for real-time operations, in *Third Symposium on Expert Systems Application to Power Systems*, Tokyo, April, pp. 370–74.
34. Okada, E., Yokoyama, E. and Sasaki, H. (1986) knowledge based system for real-time load frequency control in multi-area power system, in *Symposium on Expert Systems Application to Power Systems*, Stockholm–Helsinki, August, pp. 11-20–11-27.
35. Vadari, S.V. and Venkata, S.S. (1990) Expert load shedding: a hybrid load shedding expert system in the real-time EMS environment, in *Tenth Power Systems Computation Conference*, Austria, August, pp. 929–35.
36. Girotti, T.B., Tweed, N.B. and Houser, N.R. (1990) Real-time var control by SCADA. *IEEE Transactions on Power Systems*, **5**(1), 61–63.
37. Laik, K.C., Lee, W.J. and Chen, M.S. (1991) Design of a microcomputer-based operator assistance system for real-time voltage and reactive power correction. *IEEE Transactions on Power Systems*, **6**(2), 723–28.
38. Sun, Q.-H., Guo, X.-M., Wang, S.-M. and Xia, D.Z. (1989) A real-time expert system for substation fault diagnosis, in *Second Symposium on Expert Systems Applications to Power Systems*, Seattle USA, July, pp. 189–93.
39. Crider, C. and Hauser, M. (1990) Real time T&D applications at Virginia Power. *IEEE Computer Applications in Power*, July, 25–29.
40. Gamm, Z. (1990) New approaches to the solution of real-time problems in power systems, in *Tenth Power Systems Computation Conference*, Austria, August, pp. 1004–11.
41. Nelson, R.J., Brennan, T.F. and Engelhardt, J.S. (1989) The application of real-time monitoring and rating to HPOF pipe cable systems. *IEEE Transactions on Power Delivery*, **4**(2), 850–56.
42. Wu, F.F. (1988) Real-time network security monitoring, assessment and optimization. *EPES*, **10**(2), 83–100.
43. Mizutori, T., Takahashi, K., Maneki, Y. and Matsuki, N. (1991) rtKDL: a real-time expert shell applied to power system monitoring, in *Third Symposium on Expert Systems Application to Power Systems*, Tokyo, April, pp. 195–201.
44. Moriguchi, S., Taniguchi, T., Kunugi, M., Shimada, K. and Suzuki, K. (1989) A large-scale SCADA system with real-time knowledge based functions, in *Second Symposium on Expert Systems Applications to Power Systems*, Seattle USA, July, pp. 21–27.
45. Pavard, M., Legaud, P., Monnier, B. and Ancelin, J. (1991) Expert systems for nuclear power plants: the experience of Electricite de France, in *Third Symposium on Expert Systems Application to Power Systems*, Tokyo, April, pp. 135–38.
46. Kang, S.-J., Park, N.-S. and Kwon, Y.-R. (1991) A hybrid expert system for eddy current based inspection of steam generator tubes in nuclear power plant, in *Third Symposium on Expert Systems Application to Power Systems*, Tokyo, April, pp. 144–51.

47. Kurki, M., Majanne, Y. and Ruokonen, T. (1991) Real-time expert system for power plant diagnosis, in *Third Symposium on Expert Systems Application to Power Systems*, Tokyo, April, pp. 160–64.
48. Elzer, P., Berchers, H.W., Weisang, C. and Zinser, K. (1986) A pre-prototype of an 'intelligent' process monitoring system, in *Symposium on Expert Systems Application to Power Systems*, Stockholm–Helsinki, August, pp. 11-1–11-8.
49. Fushimi, T., Matsumura, Koi, T., Momoeda, K. and Kamiya, A. (1986) Turbine-generator on-line monitoring and diagnostic expert system, in *Symposium on Expert Systems Application to Power Systems*, Stockholm–Helsinki, August, pp. 11-14–11-19.
50. Martin, T.P., Glasgow, J.I., Féret, M.P. and Kelly, T. (1991) A knowledge based system for fault diagnosis in real-time engineering applications, *Database and Expert Systems Applications*, Berlin, pp. 287–92.
51. Okamoto, H., Yokoyama, A. and Sekine, Y. (1990) A real-time expert system for fault section estimation using cause-effect network, in *Tenth Power Systems Computation Conference*, Austria, August pp. 905–12.
52. Wagenhauer, M. and Brugger, H. (1986) Application of an expert system to locate earth faults in a real-time environment, in *Symposium on Expert Systems Application to Power Systems*, Stockholm–Helsinki, August, pp. 11-9–11-13.
53. Liu, Ch.Ch., Tsai, M.-S., Mes, V.N. and Hartwell, R. (1991) Alleviation of abnormal feeder operating conditions with a real-time expert system environment, in *Third Symposium on Expert Systems Application to Power Systems*, Tokyo, April, pp. 308–14.
54. Chan, E.-K. (1991) A real-time expert system integrated in a large SCADA system, in *Third Symposium on Expert Systems Application to Power Systems*, Tokyo, April, pp. 495–98.
55. Santoso, N.I. and Tan, O.T. (1990) Neural-net based real-time control of capacitors installed on distribution systems. *IEEE Transactions on Power Delivery*, 5(1), 266–72.
56. Algoud, J.P. (1991) Expert System and transformer diagnosis, in *Proceedings of the Seminar on Expert Systems and Case-Tools*, University of Lodz, September, pp. 7–15.
57. Dijk, H.E. and Kema, N.V. (1987) Exformer, an expert system for transformer fault diagnosis, in *Ninth Power Systems Computation Conference*, Portugal, August, pp. 715–21.
58. Kirtley, J.L. Jr, Hagman, W.H. *et al.* (1996) Monitoring the health of power transformers. *IEEE Computer Applications in Power*, January, 18–23.
59. Sun, Y., Liu, Ch.-Ch., Christie, R.D., Nordstrom, J., Hofmann, M., Stemler, G. and Thurein, I. (1991) RETEX (RElay Testing EXpert): an expert system for analysis of relay testing equipment, in *Transmission and Distribution Conference*, Dallas, Texas, September, pp. 376–84.
60. Hayes, R.M. (1991) AI in design: searching for differential relay protection. *IEEE Computer Applications in Power*, **4**(3), 21–26.
61. Lee, S.J. and Liu, C.-C. (1996) Intelligent approach to coordination identification in distance relaying, in *Proceedings of the International Conference on Intelligent System Applications to Power Systems (ISAP'96)*, pp. 62–67.
62. Lundgaard, L.E., Tangen, G., Skyberg, B. and Faugstad, K. (1992) Acoustic diagnoses of GIS: field experience and development of expert systems. *IEEE Transactions on Power Delivery*, **7**(1), 287–94.
63. Booth, C., McDonald, J.R., Laycock, W.J. and Bennett, A. (1995) Distributed intelligence within power systems: the future for substation control and management systems. *CIGRE Symposium on Integrated Control and Communication Systems*, Helsinki, paper 500–06.

64. Fauquembergue, P., Kezunovic, M. *et al.* (1995) Intelligent system applications to protections, control and monitoring within substations, in *CIGRE Symposium on Integrated Control and Communication Systems*, Helsinki, paper 600–06.
65. Waterman, D.A. (1986) *A Guide to Expert Systems*, Addison-Wesley Publ. Co.

10

Future trends in intelligent systems in power systems

Alan Moyes

10.1 INTRODUCTION

Advanced information technology, including artificial intelligence techniques, will assume increasing importance in the design, manufacture, operation, monitoring and control of power plant and systems. As has been demonstrated previously in this book, intelligent knowledge based systems (IKBSs) have the notable feature of rendering scarce expertise continuously available at a range of locations. Furthermore, advances in communications technology have allowed, for example, plant performance data to be rapidly disseminated and used in a range of strategic and operational decision making. Moreover, novel and sophisticated measurement devices are making more data available more reliably. The processing power to support the realistic application of these technologies in power systems becomes ever cheaper. This chapter considers the impact of some of these developments on future trends in IKBSs in power systems.

10.2 THE IMPORTANCE OF KNOWLEDGE

A common justification for the development of IKBSs is that they are a means of collating and effectively applying scarce expertise. The scarcity of such expertise makes it expensive to maintain, and should 'experts' leave an organization, vital knowledge about its plant and processes can be lost. Indeed, not only is ownership of key knowledge important but so is the capacity to evaluate this knowledge, to discover 'new' knowledge of commercial and technical importance, and to exploit this knowledge as effectively as possible.

While the requirement for knowledge grows, key questions remain: Which knowledge? Where can knowledge be found? How should knowledge be characterized? What can this knowledge be used for? The emergence of useful answers to these questions have wide implications for the way in which IKBSs are designed and constructed. Moreover, as will be discussed in this section, new IKBS techniques have emerged, using diverse types of knowledge, which provide scope for additional and enhanced IKBS functionality.

10.2.1 Knowledge acquisition

Knowledge acquisition is a key component of IKBS development. Indeed, knowledge acquisition, the process of eliciting and analysing the 'expertise' in a given application area, is the main activity which clearly differentiates the development of IKBSs from the development of conventional software. However, it is often viewed as perhaps the most onerous, time-consuming and difficult part of the IKBS development process. The phrase 'knowledge acquisition bottleneck' [1] was coined to describe the apparently detrimental influence of this activity in many IKBS developments. But while software engineering was developing methodologies to assist in software project management and developing techniques for requirements analysis, design and specification, IKBS development tended to use a relatively unstructured 'rapid prototyping' approach [2].

The 'rapid prototyping' approach was based on a view of knowledge acquisition known as 'knowledge transfer'. In this view, the expert's knowledge existed in 'chunks', effectively nuggets of knowledge, which could be mined, represented as production rules, and hence seamlessly transferred into a rule base. By using these within an IKBS shell, incremental prototypes could be developed which could be tested and reviewed by the experts until an acceptable version was created. While the knowledge transfer view provided a model for IKBS development, it had some important flaws which helped to create the 'knowledge acquisition bottleneck'.

Knowledge acquisition, like any other complex engineering activity, requires careful management to ensure that it stays within projected timescales and delivers reasonable results. In the approach described, the expert's knowledge, acquired from a number of knowledge elicitation techniques, is encoded into an IKBS tool. It is unlikely that a non-IKBS literate engineer has either the time or the skill to perform a detailed analysis of the sufficiency and validity of the knowledge within the developed IKBS. This basic lack of feedback and communication between the expert(s) and the knowledge acquisition team in turn hampered the knowledge acquisition process because, firstly, of the lack of a 'shared language' between the two sides which would allow effective communication of relevant concepts, and secondly, without appropriate tools and techniques it is hard to evaluate progress. Therefore, projects often became frustrated through a lack of focus and direction [3].

However, a new direction in knowledge acquisition has emerged which has the potential to alleviate many of the difficulties of the 'rapid prototyping' approach. 'Knowledge modelling' developed from work on the structures used to represent knowledge. Therefore, an important aspect of knowledge modelling is analytical, identifying 'which' type of knowledge to acquire and representing it appropriately. Furthermore, knowledge modelling is also inherently a model-building activity. Thus, knowledge acquisition can become the shared process of experts and knowledge acquisition team working towards the construction of an agreed knowledge model. Inherent in this process is validation, and hence evaluation, of the acquired knowledge, which can assist in the identification of 'gaps' in the knowledge. A further advantage of knowledge modelling is that the knowledge model provides a framework for the integration of diverse sources of engineering knowledge as these reside in textbooks, documentation, reports of incidents and so on. From experience, the development of knowledge models is a useful means of generating a shared 'language' as the terms used in the models have to be defined and agreed. Thus, knowledge acquisition can be structured and focused, which provides a more effective framework for project management [4].

A number of knowledge modelling approaches have been developed including KADS [5], KEATS [6], KRITON [7] and KEW [8]. The potential of the knowledge modelling approach is a move away from IKBS prototyping and towards structured systems specification and knowledge archiving. Thus the purpose of knowledge acquisition is, arguably, to create a structured archive containing consistent descriptions of the expertise relating to an engineering task or function [4]. Knowledge archiving fulfils the criteria for capturing and retaining valuable expertise, while presenting it in an accessible form free of software jargon. The contents of a knowledge archive offer potential re-use of knowledge, for example for training or reference purposes. In addition, the knowledge archive supports the structured specification of an intelligent system for the engineering function studied. Such a specification assists in the controlled development of a subsequent IKBS.

Structured approaches to knowledge acquisition are likely to become more prevalent as the need for effective application and management of corporate knowledge resources increases. Such approaches can support efficient knowledge acquisition and, moreover, have the potential to facilitate a range of applications for acquired knowledge.

10.2.2 Model-based reasoning

Knowledge need not be solely confined to the minds of the 'experts'. Indeed, there may be models available which describe engineering plant function, for example, models derived from design studies. Model-based reasoning (MBR) [9,10] is a computational technique which can use such models to provide IKBS functions.

MBR uses models of expected behaviour to simulate how a device should be functioning. A key feature of MBR is the comparison between observed values and those values predicted by the model of expected behaviour. A resulting discrepancy between actually observed and predicted values implies the device modelled has malfunctioned.

An advantage of model-based techniques over rule-based systems is that models of expected behaviour can be derived from a design description of the device, minimizing the requirement for knowledge acquisition. The models of expected behaviour used in MBR can be mathematical, logical, physical or a combination of these or other types. Moreover, models can be structured hierarchically to provide a range of levels of detail in generated conclusions. However, as the accuracy of MBR depends on the quality of the models used, substantial efforts may need to be directed at deriving and validating appropriate models of expected behaviour.

The MBR mechanism involves initially generating a set of hypotheses on the basis of a symptom. Subsequently, the derived set of hypotheses is reduced to a candidate set which covers all of the observed behaviour of the device. Finally, candidates are selected as explanations of the discrepancy between observed and expected behaviour on the basis of additional discriminating information. This approach differs from traditional rule-based reasoning as it assesses the possible explanations for an observed symptom, in terms of the model, rather than the probable causes stored in the rule base. Therefore, MBR offers a potentially more comprehensive and detailed diagnosis than a traditional IKBS. However, this is at the price of potentially considerable computational complexity for large or complex systems.

MBR was initially applied to troubleshooting logic circuits. Initial research in the field of power systems is showing promising results in the application of MBR to power systems problems [11–13]. One trend emerging is the use of focusing techniques to contain the complexity of MBR while retaining its explanatory strengths. Therefore, the future of the application of MBR in power systems may be in 'hybrid systems', which would use MBR together with other IKBSs and related techniques to provide flexible problem-solving capabilities.

10.2.3 Case-based reasoning

Case-based reasoning (CBR) [14] is a computational technique which uses solutions to previous similar problems to solve current problems. It is argued that this approach to problem solving closely matches the way in which human experts tackle problems. Moreover, CBR is a learning technique: as new cases become available, so the system is able to automatically use this new information in problem solving.

A CBR system uses a database made up of 'cases'. Each case is a parameterized description of the historical entities, such as fault events or design descriptions, which are the subject of the CBR system. An advantage of CBR is that such cases can be derived by hand, from archived paper reports, from appropriate databases or from documentation. Therefore, there is less requirement in CBR for knowledge acquisition. However, effort is often required to develop an effective indexing scheme which allows all cases to be characterized.

CBR operates by assessing the similarity of the currently observed case to cases resident in the 'case base'. A measure of similarity is used to discriminate between potentially matching cases. Therefore, rather than explaining the solution to a problem in terms of chains of inference, CBR systems can incorporate detailed descriptions of a case, together with relevant supporting information, as a means of justifying why a conclusion was reached.

The use of CBR to provide enhanced explanations for turbine generator fault diagnosis is described by Georgin *et al.* [15]. In this application, a CBR module was developed to augment the conclusions generated by a knowledge based system, DIVA [16], which were poorly understood by plant operators. The use of the CBR module provided a means of explaining the detected incident in more detail and reminding the operators of previous similar incidents. Applications of CBR in a range of fields are described in reference 17.

Where suitable cases exist, CBR provides the option of rapidly developing an IKBS. However, it remains to be seen how many cases, particularly in complex engineering domains, are sufficient to realize an IKBS capable of providing sensible conclusions for the range of problems which might occur. Therefore, CBR systems are likely to remain as part of hybrid systems, augmenting the capabilities of additional IKBS modules.

10.2.4 Data mining

Developments in computer systems within power utilities have resulted in the availability, collection and storage of large quantities of technical and business data.

However, once this data has been used for its primary purpose, it often remains unused, archived in large computer databases. As this data may reflect elements of a utility's key activities over an extended time period, important information relating to, for example, plant performance or critical business processes might be contained within such databases. Data mining [18] (sometimes referred to as 'knowledge discovery in databases') refers to a set of techniques which can unlock the information within corporate databases. The knowledge gained from the application of data-mining techniques could then be used to, say, improve plant utilization and maintenance or target customers more effectively.

Techniques used in data-mining applications include statistical techniques such as regression analysis, CBR, rule induction, artificial neural networks, or a combination of these. The development of flexible data-mining tools has resulted in wide application in industrial, commercial and public sector applications [19]. Within power systems these techniques have been applied to a range of problems. An important area of application is the use of data-mining techniques to assess plant performance over time, particularly as this impacts on potentially financially severe maintenance and replacement decisions. It is likely that the application of data-mining techniques will become more prevalent as utilities realize the potential value of the data stored within corporate databases.

10.3 INTEGRATED SOLUTIONS FOR COMPLEX PROBLEMS

In the preceding chapters, IKBS technology has been applied to a range of power systems problems with varying degrees of success. As has been discussed in this chapter, new IKBS techniques are emerging which promise to broaden and deepen the scope and power of IKBS applications. Initial research suggests that the new approaches outlined may have a significant role to play in future IKBS applications in power systems [20].

Power systems problems are complex problems. This section considers how IKBSs might be used as a key element in tackling more ambitious, real-world power systems problems within corporate information technology environments which are becoming prevalent in utilities world-wide. Such problems might include automation of electricity distribution networks, or optimization of power station performance.

10.3.1 Control and monitoring architecture

Figure 10.1 describes the control and information architecture in which IKBS applications will be used. A feature of this architecture is that it has three plant control and monitoring layers which interface upwards

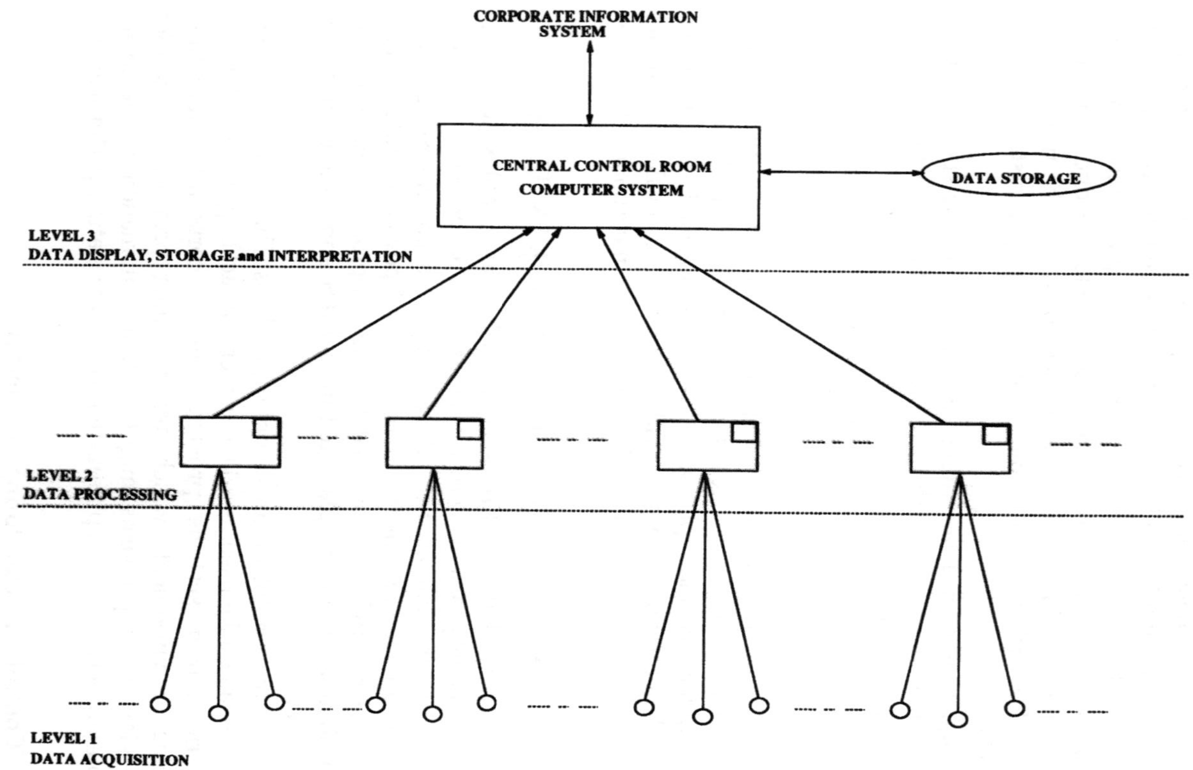

Figure 10.1 Power system plant monitoring framework.

to corporate information systems. Therefore, the operation and use of engineering plant is becoming more closely coupled to the achievement of commercial objectives, both short- and long term. This interface might allow, for example, economic requirements to directly influence plant performance targets, and the subsequent configuration and modification of control system parameters.

Level 1 in the hierarchy is the sensor/actuator level. The application of IKBSs requires reliable data from plant sensors. Increasingly, efforts are being directed towards developing 'intelligent sensors' which can perform on-board processing functions such as self-checking, and calculations of physical parameters from others measured. Moreover, this has been assisted by the development of novel sensing techniques (such as optical sensing) using novel materials, which incorporate additional physical integrity into plant measurement systems. A further development is, where groups of related sensors exist, these should be able to cross-corroborate their results so as to perform a group-wide test of sensor validity, thus isolating a faulty sensor. These developments have been assisted by the development of high-integrity, high-bandwidth communications links which support the additional data interchange.

At level 2, elementary processing is applied to data acquired from the level 1 sensor configuration. The applied processing might typically include signal conditioning, statistical analysis or fast Fourier transform analysis. At this stage, acquired data is logged in plant databases. However, advances in signal processing technology are providing the possibility of extracting more from existing plant data. Techniques such as principal components analysis, clustering, higher-order statistics and artificial neural networks have the potential to provide additional information about acquired plant data.

The role of IKBSs is to provide useful interpretation of plant data and associated information so as to give plant operators effective decision support. In addition, IKBSs may convey information to management about the performance of engineering plant. These are implemented at level 3. It is unlikely that a single IKBS using a single reasoning technique will be capable of providing satisfactory performance of the range of tasks required in, for example, operating a turbine generator or a transmission network. Firstly, as has been demonstrated in the preceding chapters, a number of power systems tasks and functions exist which require different approaches to their effective performance by IKBSs. Secondly, as has been discussed in this chapter, new techniques are emerging which provide enhanced approaches to the solution of many problems. Therefore, it is likely that either IKBSs are developed, but for individual 'niche' applications, or that bigger problems will be tackled through the use of hybrid systems which attain effective performance through the combination of a range of IKBSs and analytical techniques.

Given increasing commercial and regulatory pressures on achieving operational efficiency within power systems a trend towards using IKBSs in more complex roles is likely to emerge. The next section considers the challenge of integrating IKBSs to realize such complex and powerful information systems.

10.3.2 Integrated IKBSs

The integrated IKBS will bring a range of techniques to bear on solving complex power system problems. Key to the development of such systems will be the modern knowledge acquisition techniques described earlier in this chapter. It is likely that these will provide a framework which accurately characterizes complex power system problems while indicating the computational resources necessary to facilitate their solution. But, in addition to providing more appropriate solutions to identified problems, what further advantages would integrated IKBSs have? Experience in the design, implementation and evaluation of non-trivial IKBSs for power systems applications has highlighted the following attributes as desirable in future IKBSs [21].

1. Robustness – the performance of IKBSs has been noted to degrade significantly when the IKBS is presented with a situation not explicitly contained within its knowledge base. Moreover, IKBSs in engineering systems have a requirement to cope gracefully with faulty or missing data. Typically, this would mean that nothing, or misleading answers, would be produced by the IKBS. Consequently, the IKBS will need to be able to approach a scenario from a variety of angles in order to continue producing useful information. Research is showing that using multiple models and sources of knowledge together with multiple problem-solving methods may provide a means of increasing the robustness of IKBSs because their coverage of the problem area is greater.
2. Efficiency – rather than presenting all data to the same knowledge base the theory is emerging that only necessary knowledge should be used in solving a problem. By using that knowledge which is only necessary and sufficient to solve a problem then wasted processing time is greatly reduced. This approach is being assisted by research on characterizing and implementing problem solving strategies which adapt problem-solving behaviour to specific scenarios or types of problem.
3. Re-use of diagnostic system components – the knowledge, cases and models used by a diagnostic system are themselves of value as a coherent form of documentation relating to an engineering system. The independence of these from reasoning or problem-solving mechanisms permits their use in different applications such as training,

engineering process validation and reference. Moreover, libraries of generic reasoning and problem-solving modules for use in additional applications can be developed.

4. Flexibility – the information generated by the use of IKBSs will be made available to a wide range of interested users. Therefore, an integrated IKBS would require the flexibility to produce information tailored to the needs of particular user groups under a range of conditions. Consequently, problem-solving methods and models could be selected which tailored the style and content of information produced to the requirements of different user groups.

The implementation of such systems remains an open question. While architectures have been proposed (e.g. reference 22), clearly work will be required on developing the sophisticated implementation environments required to build such systems. Of particular importance will be issues such as communications between IKBSs performing interdependent tasks, particularly with regard to sharing different types of knowledge, data and models; the physical distribution of such systems across computer networks and the potential for accessing remote libraries of functionality on demand; 'openness' and compatibility between systems, for example between IKBSs and proprietary monitoring and control protocols; the integrity and robustness of underlying operating systems and computer hardware particularly in critical environments; the management and presentation of information to users. These, and other issues, will become increasingly important as the potential of IKBSs to provide effective solutions to complex power systems problems is realized.

10.4 CONCLUSIONS

IKBSs are proving their potential as useful tools in performing a range of power systems tasks, as has been discussed throughout this book. Developments in IKBS technology offer the possibility of developing IKBSs with an increased range of functionality addressing a broader range of more complex power systems problems. Two emerging and interacting themes have been identified: the importance of knowledge; the need for integrated, multifaceted IKBSs.

The future of IKBSs is likely to be an essential component of large-scale corporate information systems, providing a range of functions, including interfacing plant issues to business issues. In this respect, it is likely that IKBSs will provide a key supporting role for a variety of organizational functions, and if developed appropriately, provide a range of services and opportunities for both the manufacturers and users of power plant and systems.

REFERENCES

1. Buchanan, B.G., Barstow, D., Bechtal, R., Bennet, J., Clancey, W., Kulikowski, C., Mitchell, T., and Waterman, D.A. (1983) Constructing an expert system, in *Building Expert Systems* (eds F. Hayes-Roth, D. Waterman and D.B. Lenat), Addison-Wesley.
2. Johannsen, G. and Alty, J.L. (1991) Knowledge engineering for industrial expert systems, *Automatica*, **27**(1), 97–114.
3. Musen, M.A. (1993) Knowledge acquisition, in *Second Generation Expert Systems* (eds J.-M. David, M.-P. Krivine, and R. Simmons), Springer-Verlag.
4. McDonald, J.R., Burt, G.M. and Moyes, A. (196) knowledge based systems for condition monitoring, in *Proceedings of Control '96*, IEE, vol. 2, pp. 1424–29.
5. Wielinga, B.J., Schrieber, A. Th. and Breuker, J.A. (1992) KADS: a modelling approach to knowledge engineering. *Knowledge Acquisition*, **4**(1), 5–53.
6. Motta, E., Rajan, T. and Einstadt, M. (1989) Knowledge acquisition in KEATS-2, in *Topics in Expert System Design* (eds Guida and Tasso), North Holland.
7. Diederich, J. and Linster, M. (1989) Knowledge based knowledge elicitation, in *Topics in Expert System Design* (eds Guida and Tasso), North Holland.
8. Terpstra, P., Heijst, V.G., Shadbolt, N. and Wielinga, B. (1993) KA Process support through generalised models, in *Second Generation Expert Systems* (eds J.-M. David, M.-P. Krivine and R. Simmons), Springer-Verlag.
9. Davis, R. and Hamscher, W. (1988) Model-based reasoning: troubleshooting, in *Exploring Artficial Intelligence: Survey Talks from the National Conferences on Artificial Intelligence* (ed. E. Shrobe).
10. Hamscher, W., Console, L. and DeKleer, J. (eds) (1992) *Readings in Model Based Diagnosis*, Morgan Kaufman.
11. McArthur, S.D.J., Dysko, A., McDonald, J.R., Bell, S.C., Mather, R. and Burt, S.M. (1996) The application of model based reasoning within a decision support system for protection engineers. *IEEE Transactions on Power Delivery*, **11**(4), 1748–54.
12. Torinelli, G., Capetta, L., Cermignani, S., Fabiano, A.S. and Schinco, R. (1994) An interval-based approach to fault diagnosis of power systems, in *Proceedings of the International Conference on Intelligent Systems Applications to Power Systems (ISAP-94)*, vol. 2, pp. 809–16.
13. Beschta, A., Dressler, O., Freitag, H., Montag, M. and Struss, P. (1993) A model-based approach to fault classification in power networks. *Intelligent Systems Engineering*, **2**(1), 3–14.
14. Ketler, K. (1993) Case-based reasoning: an introduction. *Expert Systems Applications*, **6**, 3–8.
15. Georgin, E., Bordin, F. and McDonald, J.R. (1995) Using prototypes in case based diagnosis of steam turbines, in *Case Based Reasoning: Prospects for Applications*, IEE, March.
16. David, J.-M., Krivine, J.-P. and Ricard, B. (1993) Building and maintaining a large knowledge based system from a knowledge level perspective: the DIVA experiment, in *Second Generation Expert Systems* (eds J.-M. David, J.-P. Krivine and R. Simmons), Springer-Verlag.
17. *IEE Colloquium on Case Based Reasoning: Prospects for Applications*, IEE, March 1995.
18. Fayyad, U.M., Piatesky-Shapiro, G., Smyth, P. and Uthurusamy, R. (eds) (1996) *Advances In Knowledge Discovery and Data Mining*, MIT Press.
19. *IEE Colloquium on Knowledge Discovery in Databases*, IEE, January 1995.
20. *Proceedings of the International Conference on Intelligent Systems Applications to Power Systems (ISAP-96)*, IEEE, 1996.

21. David, J.-M., Krivine, J.-P. and Simmons, R. (1993) *Second Generation Expert Systems*, Springer-Verlag.
22. McDonald, J.R., Moyes, A. and Georgin E.M. (1996) Turbine-generator monitoring: towards integrated systems, in *Proceedings of the International Conference on Intelligent Systems Applications to Power Systems (ISAP-96)*, pp. 151–55.

Index

Page references in *italics* refer to figures

Abstract data type approach 60
AC/DC substation grounding 82
AHFA 162
Alarm processing 119–52
  alarm volume 119–20
  APEX (Alarm Processing Expert System) 143–52
    details of construction 144–6
    example of operation 146–51
    performance 151–2
  definition 119
  development 130–43
    identification of approach and IKBS techniques 137–8
    identification of existing data and knowledge 135
    identification of knowledge and data requirements for the IKBS 133–5
    identification of language or tools for IKBS construction 138
    IKBS construction 140–1
    installation and integration 142–3
    knowledge acquisition 135–7
    knowledge base design 139–40
    overall methodology 130, *131*
    prototyping 141
    setting initial targets 132–3
    testing 141–2
    user interface design 140
  history 120–7
    advanced alarm processing techniques 121–3
    advanced techniques 121–3
    fault diagnosis 122–3, *123*
    message prioritization 121
    message routing 120–1
    review of previous work 123–7
  practical problems 127–30
    inconstant arrival rate of alarms 129–30
    limited scope of telemetry 128
    missing alarm messages 128
    multiple simultaneous events 130
    speed of operation 127–8
    unreliable time ordering of alarms 129
APEX (Alarm Processing Expert System) 143–52
  details of construction 144–6
    knowledge base structure 144
    structure of network topology database 145–6
  example of operation 146–51
  performance 151–2
Artificial intelligence (AI)
  current trends 7
  history 5–7
ASDEP 80–1
ASTA system 158–60
Automatic generation control (AGC) 29

Automatic system trouble analysis (ASTA) system 158–60

Backwards chaining 9
Bellman's principle of optimality 50

C 14
  in alarm processing systems 125, 138, 143, 152
  in forecasting and planning system 60
C++ 14
Cable system design 82
CADUCEUS 17
Capacitor placement in distribution feeders 81
Certainty factors, theory of 178
Common-sense knowledge problem 6
Composite model 41
Computer-aided design (CAD) 29
Computer-aided manufacture (CAM) 29
Computer-assisted intelligent storm outage evaluator (CAISOE) 163
Computers, history 5–7
Condition monitoring 99–115
  applications 106–7
  definition 102–5
  IKBSs in 105–6
  successful applications 107–15
    GEMS (Generator Expert Monitoring System) 107, 110–11
    GenAID 107, 108–10
    MOTORMON 107–8, 114–15
    TOGA (Transformer On-Line Gas Analysis) 107, 111–14
    TURBOMAC 107, 108
  types of electrical machines 100–1
  types of network failures 101
Control and information architecture 211–14
Conventional systems 15
CRAFT (customer restoration and fault testing) 165
Curse of dimensionality 51

Decision trees 72
Demand forecasting 70–1
Demster-Shafer theory 178
DENDRAL 6, 13, 17
Design and planning of electrical plant and systems
  applications to design of specific subsystems or equipment 82–3
  applications to large systems 79–80
  applications to systems of limited dimensions 80–1
  benefits of using IKBS 77–8, 95–6
  electrical auxiliary system design in a power station 83–95
    design methodology 84–6, *85*
    IKBS approach 87–95
    problem formulation 83–7
  general overview 75–6
  need for flexibility 75
  reasons for application of IKBS to design 76–8
  reasons for application of IKBS to planning 76–7
Designer friendly approach 63
Digital relay protection 29
Direct load control (DLC) 29
Distributed problem solving (DPS) 126
  Distribution network planning, applications of IKBS 60–4
  connection of new customers 62–4
  ENH/HV substation location and HV network design 61–2
  interactive load flow 65
  support for transient stability studies 66–70
  support to load flow calculations 65–6
Distribution subsystems, applications to 81
DSF (diagnosis of system faults) 160
Duval's triangle 112, *112*
Dynamic programming 37–8, 50–1

EKA 166–7
Electrical power systems
  equations used in engineering and 32–3
  hardware and software 29–30
  history of computer applications 29
  main characteristics 27–8
  problems in operation 28

Emergency control and restoration 195, 196–8
EMYCIN 13, 14
Endorsements, theory of 178
Energy control system (ECS) 29
Energy management systems (EMS) 10, 29
  expected life 30
  processor speeds 30
ENIAC 5
Environmental protection 28
ESAPD (Expert System for Alarm Processing and Diagnosis) 127
Event diagnosis *see* fault diagnosis
Expert system, definition 8
Explanation system, definition 10

Fault diagnosis 28
  basis for diagnosis 156–7
  events 155–6
  IKBS-based 158
    applications where SCDA data is lacking 163
    automatic system trouble analysis (ASTA) system 158–9
    CRAFT (customer restoration and fault testing) 165
    distributed approach 160–1
    EKA 166–7
    fault diagnosis using object-oriented features 161–2
    IKBS based on ASTA ideas 159–60
    localized fault diagnosis 162
    SEPT 165–6
    system for distribution feeders 164
    system suited to parallel implementation 164
    SUBES 167
  reasons for 156, 157–8
  see also RESPONDD
FAUST 161–2
First-generation systems 15
Flavours 161
Forecasting and planning, systems
  evolution of techniques for 37–8
  forecasting tasks 36–7
  IKBSs and 38–44
    algorithmic, human and expert system capabilities 39–40
    assessment of solution techniques 43
    characteristics of tasks 40–2
    frequency of study 42
    retention of expertise 42–4
  planning tasks 35–6
  see also Generation expansion planning
Fortran in forecasting and planning systems 60
Forward chaining 9
Fourier analysis 103
Fuzzy systems 7, 49, 72
  attributes 52–3, *53*
  judgement templates 52, *52*
  role of 51–3

GEMS (Generator Expert Monitoring System) 107, 110–11
GenAID 107, 108–10
Generation capacity 40
Generation expansion planning
  combined dynamic programming, IKBS and fuzzy sets approach 49–55
    dynamic programming process 50–1, 54–5
    example 53–4
    methodology 49–53
    objective 49
    role of fuzzy logic 51–3
    role of IKBS 51
    structure 50
  cumulative generation cost curve 45, *46*
  decision tree and IKBS model 46–8
    data 47
    example 48
    implementation 48
    model 47–8
    objective 47
  demand duration curve *45*
  proposed and actual applications 45–55
Generation location 40–1
Generation mix 40
Genetic algorithm techniques 7, 72, 126

GESTAL application 162
GOLDWORK II 82
Graphical map system (GMS) 63–4

HEARSAY 13
Hyperbolic quasilinear partial differential equations 32

IKBS
applications 14
architecture 9–10
definition 8, 42–3
generations 15
history 13–15
implementation cycle 11
integrated 214–15
opportunities and bottlenecks in development 16–17
spread of 30–2
structure *10*
in technology 15–16
terminology 8–9
tools 13–14
Inference engine, definition 9
Integrated IKBS 214–15
Intelligent knowledge based systems, *see* IKBS
Intelligent tutoring systems (ITS) 78

KADS 208
KEATS 208
KEW 208
Knowledge, importance of 206–11
case based reasoning 210
data mining 210–11
knowledge acquisition 207–8
model based reasoning 209
Knowledge acquisition bottleneck 207
Knowledge base (KB), definition 9
Knowledge based support for voltage collapse detection and prevention (KBVCDP) 196
Knowledge discovery in databases 211
Knowledge engineer, definition 10
Knowledge modelling 208
Knowledge transfer 207
Knowledge update, definition 9–10
KRITON 208

Large cage-rotor induction motors 107
Likelihood ratios, theory of 178
Linear programming 37
LISP 14
in alarm processing systems 138, 139, 152
in design protection support system 81
in electrical plant design 81, 87
Load control 28
Load flow systems 65–6
Load management control system (LMCS) 29

M-P unit 6
MACSYMA 13
Management systems 200
MOTORMON 107–8, 114–15
MYCIN 13, 14

Neural networks 7, 72
history 5–6
NEXPERT 87, 88, 89, 91–2
NEXPERT Object 57, 200
NSP (Northern States Power) Intelligent Alarm Processor 124, 125
Nuclear industry, alarm processing 127

Off-line systems 15
On-line systems 15
OPS5 13, 60
Optimization techniques 37–8

Pascal
in alarm processing systems 138
in forecasting system 48
Pattern recognition 72
Perceptron 6
Planning *see* Design and planning of electrical plant and systems
POP11 161
Possibility theory 178
Power transformers 106
Production costing algorithm 45
Project frame (PF) 50
Prolog 14
in alarm processing systems 138, 139, 152
in distribution substation system 81

in fault diagnosis 170–1, 191
in forecasting and planning system 60, 61, 64
PROSPECTOR 13, 17
Protection coordination in industrial systems 82–3
Protection relays, testing of 200

R1 17
Rapid prototyping approach 207, 208
Reactive management and voltage monitoring and control (RVMC) 195–6
Real-time processing 198–9
Relay protection 28
Research and development 28
power systems 44–5
RESPONDD 167–87
dealing with uncertainty 177–9
problem factors 178
relative likelihood 178–9
source of uncertainty 177–8
demonstration of developed system 187–90
case study 187–8
hypothesizer diagnosis 189–90
hypothesizer gut reaction 174
hypothesizer strategy 173–4
hypothesizing strategies for central diagnosis 174–7
derived hypothesis and scenario generation 176–7
original hypothesis and scenario generation 175–6
subsequent testing 176
initial prototype 171–3
assessment 173
central diagnosis 170–1, 172–3
gut reaction 170, 171–2
modelling knowledge 184–7
faults on plant 184–5
plant internal links 185–6
protection and switchgear activity 186–7
protection detection logic 186
network data requirements 179–80
dynamic data requirements 179–80
static data requirements 180
structure of power system data 180
qualitative simulation 181–4
explanation/understanding facilities 182–3
fault path finding 183–4
function 181
qualitative simulation knowledge 181–2
reason for being qualitative 181
structure of the diagnosis
mechanics 170–1
results 168–70
RETEX 200
RI 13
RULE-MASTER 82

SAA 125–6
SCADA 120, 121
Scoring techniques 178
Second-generation systems 15
SELF 66, 79–80
SEPT 165–6
Shallow reasoning systems 15
Shell, definition 14
Simulated annealing 72
SINTIA II 58–60
application of IKBS 59–60
interactive planning 58–9
linearized power flows and new circuits for addition 59
objective 58
SINTRA 58, 60
Smalltalk-80 162
SOCCS (System Operation Computer Control System) Alarm Advisor (SAA) 125–6
State frame (SF) 50
Steady-state analysis 32
SUBES 167
Subjective probability theory 178
Substations, testing of 200
Subtransmission and distribution network extensions 41
Supervisory control and data acquisition (SCADA) systems 10, 29
System analysis 41–2

Testing of equipment 199–200

Three Mile Island nuclear power accident 30
TOGA (Transformer On-Line Gas Analysis) 107, 111–14
Training 28
TRANSEPT 56–8
  costing 57
  implementation 57–8
  learning 57
  objective and basic approach 56
  overall structure 56, *56*
  search strategies 57
  simulation control 57
  system design (AC system) 57
Transformers, testing of 199–200
Transient stability studies 66–70
  knowledge based support system for power system analysis 70
  transient stability IKBS 66–9, *68*
Transition control rules 50
Transmission and distribution plant standardization 41
Transmission network extensions 41
Transmission network planning
  long-term planning 58–60
  preliminary design 55, 56–8
Transmission systems 107
Turbine generators106
TURBOMAC 107, 108

Vibration fingerprint 103–4
Voltage control 28
Voltage/reactive power (VAR) control 195

Window rules 50

MIX
Papier aus verantwortungsvollen Quellen
Paper from responsible sources
FSC® C105338

If you have any concerns about our products,
you can contact us on
ProductSafety@springernature.com

In case Publisher is established outside the EU,
the EU authorized representative is:
Springer Nature Customer Service Center GmbH
Europaplatz 3, 69115 Heidelberg, Germany

Printed by Libri Plureos GmbH
in Hamburg, Germany